QG: The strange theory of Space, Time and Matter

Beyond Einstein, Volume 6

Balungi Francis

Published by dePhysique Hub, 2018.

Also by Balungi Francis

Beyond Einstein
Quantum Gravity in a Nutshell1
Solutions to the Unsolved Physics Problems
Mathematical Foundation of the Quantum Theory of Gravity
A New Approach to Quantum Gravity
Balungi's Approach to Quantum Gravity
QG: The strange theory of Space,Time and Matter
The Holy Grail of Modern Physics
Fifty Formulas that Changed the World
Quantum Gravity in a Nutshell1 Second Edition
What is Real?:Space Time Singularities or Quantum Black Holes?Dark
Matter or Planck Mass Particles? General Relativity or Quantum Gravity?
Volume or Area Entropy Law?
The Holy Grail of Modern Physics
Brief Solutions to the Big Problems in Physics, Astrophysics and Cosmology

Brief Solutions to the Big Problems
Brief Solutions to the Big Problems

Pursuing a Nobel Prize
Serious Scientific Answers to Millennium Physics Questions

Using Geographical Information Systems to Create an Agroclimatic Zone
map for Soroti District

Think Physics
Proof of the Proton Radius
Emergence of Gravity
On the Deflection of Light in the Sun's Gravitational Field
Reinventing Gravity

Standalone
Using Gis to Create an Agroclimatic Zone map for Soroti Distric
Expecting
Quantum Gravity in a Nutshell 2
Balungi's Guide to a Healthy Pregnancy
Prove Physics
The Origin of Gravity and the Laws of Physics
Derivation of Newton's Law of Gravitation
When Gravity Breaks Down

Table of Contents

QG: The Strange Theory of Space Time and Matter
BALUNGI FRANCIS

Visionary School of Quantum Gravity

Independent Publishers

Kampala, Uganda, East Africa

+256 (0) 777 105 605

Email: bfrancis@cedat.mak.ac.ug, balungif@gmail.com

First Book Edition: 2018

Cover: **rebecacovers**

Dedication

To my wife for his constant feedback throughout and many long hours of editing, and friends who offered their time and comments along the way, To **Carlo Rovelli, Neil deGrasse Tyson, Sabine Hossenfielder, Francesca Vidotto, Lee Smolin, Abhas Mitra** and **Rima Meta** for always being controversial and their love to fight for the foundations and the future of physics. Lastly, to the family and friends of the late **Albert Einstein** and **Stephen W. Hawking** who have rated my works in equal proportion to the works of these two great men. Thank you

Preface

There is a need for a book on a Quantum Theory of Gravity that is not directed at specialists but, rather, sets out the concepts underlying this subject for a broader scientific audience and conveys joy in their beauty. Balungi has written with this goal in mind, and has succeeded admirably. This wonderful and exciting book is optimal for physics graduate students and researchers. The physical explanations are exceedingly well written and integrated with formulas. Quantum Gravity is the next big thing and this book will help the reader understand and use the theory.

Author's Note

Our search for ultimate understanding—the Quantum Theory of Gravity—has long been the quest of such great scientists as Aristotle, Newton, Einstein, Hawking and many others, and is expected to transform science, providing clarity and understanding that is unknown today, ideally via one single overlooked principle in nature. So far, this quest has produced theories such as Special Relativity, General Relativity and Quantum Mechanics, and such recent proposals as "Dark Matter" and "Dark Energy" in cosmology. Yet these all suffer serious internal problems and compatibility issues with each other, introducing even more questions, mysteries and paradoxes—and often even violations of our laws of physics upon closer examination. As a result, the Quantum Theory of Gravity continues to elude us, leaving a fractured and divided scientific community with no clear direction forward. This has also resulted in the mathematisation of physics which has resulted in the reduction of the cosmos to a mathematical entity, which has not only confused physicists but accounts for their worst and most distracting assertions. This book makes a first case for the latter, with clear discussions exposing the flaws in the above concepts and more, while stepping back to take a good look at the scientific legacy we have inherited.

We are probably asking the wrong questions at the moment, nevertheless it is impossible to resist the temptation to try. After all, the other fundamental forces – except gravity – fit very neatly with quantum mechanics.

Physics is an entity and therefore requires only one subject to describe it fully and this subject is quantum gravity.

Balungi Francis 2018

Chapter 1.Solving Quantum Gravity

To the intra-atomic movement of electrons, atoms would have to radiate not only electromagnetic but also gravitational energy if only in tiny amounts. As this is hardly true in nature, it appears that quantum theory would have to modify not only Maxwellian electrodynamics, but also the new theory of gravitation.

Albert Einstein

The development of a quantum theory of gravity began in 1899 with Max Planck's formulation of "Planck scales" of mass, time, and length. During this period, the theories of quantum mechanics, quantum field theory and general relativity had not yet been developed. This means that Planck himself had no idea about what he had just developed-behind the Black board. Planck was not aware of quantum gravity and what it would mean for physicists. But he had just coined in formula one of the starting point for the holy grail of physics.

After P.Bridgman's disapproval of Planck's units in 1922, Albert Einstein having published the General Relativity theory, a few months after its publication he noted that "to the intra-atomic movement of electrons, atoms would have to radiate not only electromagnetic but also gravitational energy if only in tiny amounts, as this is hardly true in nature, it appears that quantum theory would have to modify not only Maxwellian electrodynamics, but also the new theory of gravitation". This showed Einstein's interest in the unification of Planck's quantum theory with his newly developed theory of Gravitation.

Then in 1933 came Bronstein's cGh-plan as we know it today. In his plan he argued a need for Quantum Gravity. In his own words he stated: "After the relativistic quantum theory is created, the task will be to develop the next part of our scheme that is, to unify quantum theory (h), special relativity (c) and the theory of gravitation (G) into a single theory". Thus the theory of quantum gravity is expected to be able to provide a satisfactory description of the microstructure of space time at the so called Planck scales, at which all fundamental constants of the ingredient theories, c (speed of light), h (Planck constant) and G (Newton's constant), come together to form units of mass, length and time.

Therefore the need for the theory of quantum gravity is crucial in understanding nature, from the smallest to the biggest particle ever known in the universe. For example, "we can describe the behavior of flowing water with the long- known classical theory of hydrodynamics, but if we advance to smaller and smaller scales and eventually come across individual atoms, it no longer applies. Then we need quantum physics just as a liquid consists of atoms" Daniel Oriti in this case imagines space to be made up of tiny cells or atoms of space and a new theory of quantum gravity is required to describe them fully.

The aim of this book is to develop a theory capable of explaining the quantum behavior of the gravitational fields and thereafter explain the problems involving a combination of very high energy and very small dimensions of space such as, the behavior of Black holes and the study of the properties of the early universe.

For us to solve the problem of quantum gravity (QG) we need to address and understand in detail the situations where the general theory of relativity (GR) fails. Below I outline briefly where GR breaks down;

(i) General relativity fails to account for dark matter.

(ii) General relativity fails to explain details near or beyond space-time singularities. That is, for high or infinite densities where matter is enclosed in a very small volume of space. Abhay Ashtekar says that; when you reach the singularity in general relativity, physics just stops, the equations break down

(iii) General relativity also fails to be quantized.

The demand for consistency between a quantum description of matter and a geometric description of spacetime, as well as the appearance of singularities (where curvature length scales become microscopic), indicate the need for a full theory of quantum gravity. For example; for a full description of the interior of black holes, and of the very early universe, a theory is required in which gravity and the associated geometry of space-time are described in the language of quantum physics. Despite major efforts, no complete and consistent theory of quantum gravity is currently known, even though a number of promising candidates exist.

The first step towards the development of a quantum theory of gravity lies in studying the kind of physics behind white dwarfs, neutron stars and black holes which are born when normal stars die. White dwarfs are supported by the pressure of degenerate electrons, Neutron stars are supported largely by the pressure of degenerate neutrons while Black holes on the other hand, are completely collapsed stars that is, stars that could not find any means to hold back the inward pull of gravity and therefore collapse to a singularity.

This section is aimed at answering questions like; i) Do objects continually collapse to a singularity or there is a limiting distance below which the very notions of space and length cease to exist?

Theorem:- A star more than three times the size of our Sun collapses in this way; the gravitational forces of the entire mass of a star overcomes the electromagnetic forces of individual atoms and so collapse inwards. If a star is massive enough it will continue to collapse creating a Black hole, where the whopping of space time is so great that nothing can escape not even light, it gets smaller and smaller. The star in fact gets denser as atoms even subatomic particles literally get crashed into smaller and smaller space, and its ending point is of course a space time singularity.

The appearance of singularities in any physical theory is an indication that either something is wrong or we need to reformulate the theory itself. Singularities are like dividing something by zero. One such theory plagued by singularities is the General theory of relativity (GR) and the problems in GR arise from trying to deal with a universe that is zero in size (infinite densities). However, quantum mechanics suggests that there may be no such thing in nature as a point in space-time, implying that space-time is always smeared out, occupying some minimum region. The minimum smeared-out volume of space-time is a profound property in any quantized theory of gravity and such an outcome lies in a widespread expectation that singularities will be resolved in a quantum theory of gravity. This implies that the study of singularities acts as a testing ground for quantum gravity.

Loop quantum gravity (LQG) suggests that singularities may not exist. LQG states that due to quantum gravity effects, there must be a minimum distance beyond which the force of gravity no longer continues to increase as the distance between the masses become shorter or alternatively that interpenetrating particle waves mask gravitational effects that would be felt at a distance. It must also be true that under the assumption of a corrected dynamical equation of LQ cosmology and brane world model, for the gravitational collapse of a perfect fluid sphere in the commoving frame, the sphere does not collapse to a singularity but instead pulsates between a maximum and minimum size, avoiding the singularity.

The resolution of classical singularities under the assumption of a maximal acceleration has been studied using canonical methods for Rindler, Schwarzschild, Reissner-Nordstrom, Kerr-Newman and Friedman-Lemaitre metrics. However in this book we use different methods to arrive at fundamental physical principles.

One of the first and crucial step towards the development of a quantum theory of gravity is the resolution of singularities that plague the Einstein General theory of Relativity. The solution has been given in part by Carlo Rovelli and Francis Viddoto in LQG but in this book we develop a different approach towards the problem as;

Let the Heisnberg Uncertainity principle be modified and written as,

$$r \times p = \frac{\hbar}{2}\alpha^n$$

Where r represents the size of a star, in this case-horizon radius, p is the momentum of a particle approaching or falling into the hole of a star, α is the coupling constant and n is positive.

From the modified uncertainty relation given above, we prove the existence of a maximal acceleration which in turn yields a bound on temperature, curvature and on the energy density in appropriate cosmological contexts, supporting the results in LQ Cosmology and for Black holes.

From the uncertainity relation we deduce the acceleration as,

$$a = \frac{c^2}{2r}\alpha^n$$

Taking , $\alpha = \frac{GM^2}{\hbar c} = \left(\frac{M}{M_{pl}}\right)^2$ (Gravitational coupling constant, where M_{pl} is Planck mass) and $r = \frac{2GM}{c^2}$ (the Schwarzichild radius), then the maximal acceleration that a particle can have when n=1/2 will be given by,

$$a = \left(\frac{c^7}{16Gh}\right)^{1/2}$$

This acceleration also imposes a general upper bound on temperature given by;

From the Unruh temperature $T = \frac{\hbar a}{2\pi c k}$, If we subsititute for the maximal acceleration we have,

$$T = \frac{1}{8\pi k}\left(\frac{c^5 h}{G}\right)^{1/2}$$

Therefore a maximal acceleration screens the classical singularity.

Since the energy density is expressed as force per unit surface area of a star. Where the force is mass times acceleration we have,

$$\rho = \frac{ma}{4\pi R^2}$$

which gives,

$$\rho = \frac{mc^2}{8\pi R^3}\alpha^n$$

On conditions that $R = \frac{2GM}{c^2}$, $\alpha = \frac{GM^2}{hc} = \left(\frac{M}{M_{pl}}\right)^2$ and n=1 we have a maximum energy density value as,

$$\rho = \frac{c^7}{\pi h (8G)^2}$$

Nature appears to enter the quantum gravity regime when the energy density of matter reaches the Planck scale. The point is that this may happen well before relevant lengths become planckian. For instance, a collapsing spatially compact universe bounces back into an expanding one. The bounce is due to a quantum gravitational repulsion which originates from the Heisenberg uncertainty, and is akin to the "force" that keeps an electron from falling into the nucleus. The bounce does not happen when the universe is of planckian size, as was previously expected; it happens when the matter energy density reaches the Planck density.

To resolve the singularities of General relativity we consider the possibility that the energy of a collapsing star and any additional energy falling into the hole could condense into a highly compressed core with density of the order of the Planck density. If this is the case, the gravitational collapse of a star does not lead to a singularity but to one additional phase in the life of a star: a quantum gravitational phase where the gravitational attraction is balanced by a quantum pressure.

For example, if we let the acceleration due to gravity of a star be in eqillibrium with the acceleration of the quantum particle falling into a hole (From modified UP), then the size of a star will be given following the derivation below,

$$\frac{GM}{R^2} = \frac{c^2}{2r}\alpha^n$$

Where R is the distance between a star and a particle.

r is the size of a star (From the uncertainty principle)

Then,

$$r = \frac{c^2 R^2}{2GM}\alpha^n = \frac{R^2}{R_s}\alpha^n$$

Where $R_s = \frac{2GM}{c^2}$ is the radius of the event Horizon of a star.

Therefore on condition that, $\frac{R^2}{R_s} = l_p$ (Planck Length) and $\alpha = \left(\frac{M}{M_{pl}}\right)^2$ we have the size of a star as,

$$r = \left(\frac{M}{M_{pl}}\right)^{2n} l_p$$

Where M is the mass of the star and n is positive.

For instance, if n = 1/6, a stellar-mass black hole would collapse to a Planck star with a size of the order of 10^{-10} centimeters. This is very small compared to the original star in fact, smaller than the atomic scale but it is still more than 30 orders of magnitude larger than the Planck length. This is the scale on which we are focusing here. The main hypothesis here is that a star so compressed would not satisfy the classical Einstein equations anymore, even if huge compared to the Planck scale. Because its energy density is already planckian(for more examples see Carlo. R & F. Vidotto 2014-Planck Stars).

In conclusion the relation $\frac{R^2}{R_s} \geq l_p$ as has been given above states that lengths beyond the Planck length are meaningless and that the singularities are resolved at the Planck scale, which in this case is another form of the uncertainty principle. The results given above therefore indicate the requirement for the modification of the Uncertainty principle.

More Examples:

(1)Singularity Resolution under the Assumption of Maximal Acceleration and Minimal length for both the Schwarzschild and Reissner-Nordstrom Black Hole

Under the assumption of $\mu = m\alpha^{1/2}$ (where α is the coupling constant), in the Caianeillo maximum acceleration model ($A_{max} = \frac{\mu c^2}{m\lambda}$) , we derive the minimum radius to which a gravitating body can collapse in the commoving frame for both the Schwarzschild and Reissner-Nordstrom Black hole.

In the context of a geometrical unification of quantum mechanics and general relativity in phase space, Caianiello was the first person to propose the existence of a maximal proper acceleration for massive particles. Caianiello was able to derive the value $A_{max} = \frac{2mc^3}{\hbar}$ **(1)** for the maximum acceleration of a particle of rest mass m from the time-energy uncertainty relation. Caianiello model was based on two assumptions; $\hbar = \lambda\mu c$ and $\mu = 2m$ **(2)** for $A_{max} = \frac{\mu^2 c^3}{mh} = \frac{ch}{m\lambda^2} = \frac{\mu c^2}{m\lambda}$ **(3)**.

Applications of Caianiello's model include cosmology, the dynamics of accelerated strings, neutrino, oscillations and the determination of a lower neutrino mass bound. There is also evidence for maximal acceleration and singularity resolution in covariant loop quantum gravity found by Rovelli and Vidotto.

In this book we propose an adhoc assumption of $\mu = m\alpha^{1/2}$ where α is the coupling constant. This differs from Caianiello's model assumption of $\mu = 2m$. Therefore the maximum acceleration(3) will be given by,

$$A_{max} = \frac{c^2}{r}\alpha^{1/2}$$ **(4)**

Where, r is the smallest possible distance between any two masses. In this book r takes values for the Schwarzschild and Reissner-Nordstrom radius.

Equation (4) given above reduces to the value $A_{max} = \frac{2mc^3}{\hbar}$ that was earlier derived by Caianiello under two conditions;

(i) When $r = \dfrac{2GM}{c^2}$ and $\alpha = 16\alpha_g^{\,2}$ for a Schwarzschild Black hole of mass M. Where α_g is the gravitational coupling constant $\dfrac{GM^2}{\hbar c}$.

(ii) When $r = \left(\dfrac{Ge^2}{4\pi\varepsilon_0 c^4}\right)^{1/2}$ and $\alpha = 4\alpha_g\alpha_e$ for a Reissner-Nordstrom Black hole. Where α_e is the electromagnetic coupling constant $\dfrac{e^2}{4\pi\varepsilon_0 \hbar c}$.

(2)Maximal Acceleration in Quantum Gravity

Reissner-Nordstrom black hole

Considering the event horizon of a Reissner-Nordstrom black hole of radius $r = \left(\dfrac{Ge^2}{4\pi\varepsilon_0 c^4}\right)^{1/2}$ and gravitational coupling $\alpha = \dfrac{GM^2}{hc}$. Then substituting in (4), the growing acceleration approaching a classical singularity in the Reissner-Nordstrom metric is bounded by the existence of a maximal acceleration of;

$$a_{max} = \frac{M}{e}\left(\frac{4\pi\varepsilon_0 c^7}{h}\right)^{1/2} \qquad (5)$$

Where e is charge on an electron, ε_0 is the permittivity of free space and $\hbar$ is the reduced Planck constant.

Schwarzschild Black hole

Considering the event horizon of a Schwarzschild black hole of radius $r = \dfrac{GM}{c^2}$ and gravitational coupling $\alpha = \dfrac{GM^2}{hc}$. Then substituting in (4), the growing acceleration approaching a classical singularity in the Schwarzschild metric is bounded by the existence of a maximal acceleration of;

$$a_{max} = \left(\frac{c^7}{Gh}\right)^{1/2} \qquad (6)$$

Minimal Radius in Quantum Gravity

Because of the equivalence principle in the case of gravitational interaction, we propose to show here that the existence of a minimal length for both a Reissner and Schwarzschild Black hole is a straight forward consequence of our maximal acceleration value (4).

In Newtonian law (center of mass system)

$$\frac{GM}{R^2} = \frac{c^2}{r}\alpha^{1/2}$$

Where, R is the radius of a Black hole (In this case the minimum radius to which a central mass will collapse)

On arranging we have,

$$R = \frac{1}{\alpha^{1/4}}\left(\frac{R_s r}{2}\right)^{1/2} \qquad (7)$$

Where R_s is the Schwarzschild radius $R_s = \dfrac{2GM}{c^2}$.

Two results are thus deduced;

1) For $r = \left(\dfrac{Ge^2}{4\pi\varepsilon_0 c^4}\right)^{1/2}$ the radius of the event horizon of a Reissner Black

hole and $\alpha = \dfrac{GM^2}{\hbar c}$, the minimum radius to which a gravitating body will collapse in a commoving frame of the Reissner-Nordstrom metric will have a value;

$$R_{min} = \left(\dfrac{\hbar e^3 G^2}{4\pi\varepsilon_0 c^7}\right)^{1/4} = \alpha_e^{1/4} l_p \quad .(8)$$

Where l_p is the Planck length and α_e is the fine structure constant $1/137$.

$R_{min} = 4.724 \times 10^{-36} m$

2) Similarly, for $r = \dfrac{GM}{c^2} = \dfrac{R_s}{2}$ the radius of the event horizon of a

Schwarzschild Black hole and $\alpha = \dfrac{GM^2}{\hbar c}$, the minimum radius to which a gravitating body will collapse in a commoving frame of the Schwarzschild metric will have a value;

$$R_{min} = (M)^{1/2}\left(\dfrac{\hbar G^3}{c^7}\right)^{1/4} \quad .(9)$$

The above derivation clearly provides evidence for the existence of a maximal acceleration and minimal length which are both expected in the theory of quantum gravity to cure strong singularities such as, big bang, big crunch, black holes etc.

Applications

(1) A Simple Derivation of The Minimum Radius of a Reissner -Nordstrom Black Hole: The Case of Accretion

Matter falling onto somebody is termed accretion. Suppose the matter is falling onto a star of mass M and radius R. Falling freely, it gains kinetic energy E_k in exchange for gravitational potential energy E_p. For a mass m falling from infinity to a distance r from the central mass M where relativistic quantum effects are taken into account, the E_k matches the E_p as

$$E_k = E_p$$

As the particle orbits closer and closer into a huge gravitational field its velocity increases up to a speed of light c, where the usual known kinetic energy formula does not apply. Instead we are forced to introduce a new formula that takes into account the gravitational coupling constant α_g as

$$E_k = \alpha_g^{1/2} mc^2 \quad (1)$$

The self gravitation force of a star of radius R and mass M is known from Newton's gravitational force formula however the potential gravitational energy of a particle m falling from infinity to a distance r from a star will differ from the usual known potential formula as

$$E_p = \left(\frac{GMm}{R^2}\right) r \quad (2)$$

Then surely,

$$\alpha_g^{1/2} mc^2 = \left(\frac{GMm}{R^2}\right) r$$

on cancelling like terms we have,

$$R = \frac{1}{\alpha_g^{1/4}} \left(\frac{R_s r}{2}\right)^{1/2} \quad (3)$$

Where, $R_s = \dfrac{2GM}{c^2}$ is the Schwarzschild radius of a gravitating body and $\alpha_g = \dfrac{GM^2}{\hbar c}$ is the gravitational coupling constant that determines the strength of the gravitational force and G is the gravitational constant.

The mass eventually hits the surface of the star and its Kinetic energy is released as heat, and appears in some form of radiation. The radius of a star can then be determined using the above formula as: For a particle at the event horizon of a Reissner-Nordstrom Black hole, $r = \left(\dfrac{Ge^2}{4\pi\varepsilon_0 c^4}\right)^{1/2}$. Where e is charge on an electron, ε_0 is the permittivity of free space and $\hbar$ is the reduced Planck constant.

The radius of this star is;

$$R = \left(\frac{\hbar e^2 G^2}{4\pi\varepsilon_0 c^7}\right)^{1/4} = 4.717444838 \times 10^{-36} m.$$

Then in terms of the Planck length we have,

$R = 0.2923 l_p$

Where l_p is the Planck length $\quad l_p = \left(\frac{\hbar G}{c^3}\right)^{1/2} = 1.6144 \times 10^{-35} m$.

Taking fourth powers on both sides of the equation we have,

$R = \alpha_e^{1/4} l_p$.

Where $\alpha_e = 1/137$ is the fine structure constant.

Therefore the above derivation implies that the radius of a Reissner-Nordstrom Black Hole is quantized in units of the Planck length and takes on only discrete units implying the quantized nature of space. In conclusion nature permits the existence of a minimum length beyond which the very notions of space and time cease to exist. I hope in my own view that this analysis will be useful for researchers involved in the field of quantum gravity and loop quantum cosmology.

(2) Evidence for Minimal Length and Singularity Resolution in Quantum Gravity

General relativity predicts two kinds of singularities; the cosmological singularity at the beginning of our universe and the singularities at the centre of black holes. However, singularities signal the breakdown of general relativity and it is generally believed that they will be removed in a more fundamental theory of quantum gravity. The resolutions of singularities have been carried out directly in the frame work of Loop quantum gravity under the assumption of a maximal acceleration using canonical methods. However, in this example, singularities are resolved under the assumption of minimal length by creating a new cosmological model for the study of the gravitational collapse of a perfect fluid sphere. Two results are deduced;(i) a commoving observer accelerating with respect to his neighbors in a Reissner-Nordstrom space-time geometry will have a horizon at a distance bounded by a minimal value limit $R_{min} = \alpha_e^{1/4} l_p$ (Where l_p is the Planck length and α_e is the fine structure constant 1/137) and (ii), a commoving observer accelerating with respect to his neighbors in a Schwarzschild space-time geometry will have a horizon at a distance bounded by a minimal value limit $\quad R_{min} = (2M)^{1/2}\left(\frac{\hbar G^3}{c^7}\right)^{1/4}$.

Therefore the generic bound on length and acceleration implies that the resolution of singularities is general and must be taken seriously.

Here we consider the gravitational collapse of a perfect fluid sphere- a gravitating body of mass M and radius R. Then for a test particle or an observer falling freely from infinity to a distance R_o from the gravitating body, the spherically symmetric solution to the Einstein field equation will be given by;

$$R^2 = \frac{1}{\alpha_g^{1/2}}\left(\frac{R_s R_0}{2}\right) \quad (1)$$

Where, $R_s = \frac{2GM}{c^2}$ is the Schwarzschild radius of a gravitating body and $\alpha_g = \frac{GM^2}{\hbar c}$ is the gravitational coupling constant that determines the strength of the gravitational force, G is the gravitational constant and c is the constant speed of light.

Therefore a commoving observer accelerating with respect to his neighbors in a given space- time geometry will have a horizon at a distance

$$R = \frac{1}{\alpha_g^{1/4}}\left(\frac{R_s R_0}{2}\right)^{1/2}$$

bounded by a minimal value limit R_{min}. Correspondingly, the growing acceleration approaching a classical singularity is bounded by the existence of a maximal acceleration

$$a = \frac{c^2}{R_o}\alpha_g^{1/2}$$

The existence of minimal length and maximum acceleration is of course something long expected in the quantum theory of gravity. Below we derive two important results for minimum radius and maximum acceleration supporting the results in loop quantum cosmology and black holes.

(i) Considering a test particle at the event horizon in the Reissner-Nordstrom metric (RN), $R_0 = R_{RN} = \left(\frac{Ge^2}{4\pi\varepsilon_0 c^4}\right)^{1/2}$. Where e is charge on an electron, ε_0 is the permittivity of free space and $\hbar$ is the reduced Planck constant. The minimum radius (size) to which a gravitating body can collapse in a commoving frame is;

$$R_{min} = \left(\frac{\hbar e^2 G^2}{4\pi\varepsilon_0 c^7}\right)^{1/4} = 4.717444838 \times 10^{-36} m$$

This also implies a maximum acceleration of $a_{max} = \frac{M}{e}\left(\frac{4\pi\varepsilon_0 c^7}{\hbar}\right)^{1/2}$. Then in terms of the Planck length we have, $R_{min} = 0.2923 l_p$ (2) Where l_p is the Planck length $l_p = \left(\frac{\hbar G}{c^3}\right)^{1/2} = 1.6144 \times 10^{-35} m$. Taking fourth powers on both sides of the equation we have, $R_{min} = \alpha_e^{1/4} l_p$ (3). Where $\alpha_e = 1/137$ the fine structure constant.

(ii) Considering a test particle at the event horizon in the Schwarzschild metric, $R_0 = R_S = \frac{2GM}{c^2}$. The minimum radius to which a gravitating body can collapse in a commoving frame is; $R_{min} = \left(\frac{4\hbar G^3 M^2}{c^7}\right)^{1/4}$. This also implies a maximal acceleration of,

$$a_{max} = \left(\frac{c^7}{4Gh}\right)^{1/2}.$$

Therefore a commoving observer accelerating with respect to his neighbors in a Reissner-Nordstrom space-time geometry will have a horizon at a distance bounded by a minimal value limit $R_{min} = \alpha_e^{1/4} l_p$. Where l_p is the Planck length and α_e is the fine structure constant 1/137. Correspondingly, the growing acceleration approaching a classical singularity in this metric is bounded by the existence of a maximal acceleration $a_{max} = \frac{M}{e}\left(\frac{4\pi\varepsilon_0 c^7}{\hbar}\right)^{1/2}$ where M is mass.

Also, a commoving observer accelerating with respect to his neighbors in a Schwarzschild space-time geometry will have a horizon at a distance bounded by a minimal value limit $R_{min} = (2M)^{1/2}\left(\frac{\hbar G^3}{c^7}\right)^{1/4}$. Correspondingly, the growing acceleration approaching a classical singularity in this metric is bounded by the existence of a maximal acceleration $a_{max} = \left(\frac{c^7}{4Gh}\right)^{1/2}$.

It has been deduced in (i) above that, the resolution of singularities occurs as a result of a fundamental discreteness of space. This is based on the fact that the minimum radius or size is proportional to the Planck length l_p (2). This is one of the promising results of this essay. The presence of l_p implies a discreteness of space or length spectra which is manifested by the presence of the fine structure constant (3).However, in (ii) singularities are avoided in a limit $M = M_p$ (Planck mass), by imposing a minimum length $R_{min} = 2^{1/2}l_p$. Therefore the generic bound on length and acceleration implies that the resolution of singularities is general and must be taken seriously.

Unlike other models, the cosmological model (1) created in this example directly predicts a limit on the length and acceleration, thus providing evidence for the resolution of the classical singularity. The derivation in (i) is unique in that the value of the fine structure constant comes out as a direct result of the theory, which has never been witnessed in any promising theory of quantum gravity, not even in LQG or string theory.

Remark: In a more general form we can express (1) as, $R^2 = \frac{1}{\alpha_g^s}\left(\frac{R_s R_o}{2}\right)$, where s= 0,1,2,3,......,1/2, Such that when s=0, $R = \left(\frac{R_s R_o}{2}\right)^{1/2}$. What name should be given to s is left for the reader to decide. However we can denote s as an extra dimension number.

We have clearly modified the geometry of Rindler space by the introduction of the coupling $\alpha^{1/2}$ into the formula for acceleration. We have witnessed that the presence of $\alpha^{1/2}$ into the formula for acceleration leads to an exact evidence for the existence of the maximal acceleration and minimal length for both the Reissner-Nordstrom and Schwarzschild black holes in quantum gravity. The split horizon in a Rindler wedge at a distance R= c^2/a for the acceleration a has been modified here, hope you have witnessed how $\alpha^{1/2}$ changes all of this. This implies that there is some fundamental limitation on how much acceleration a particle could experience based on the strong-field behavior of the fundamental force causing it.

Results of the maximal acceleration and minimal length for the Reissner Black hole have not been derived anywhere in literature. These clearly impose a general bound on acceleration and length (in Reissner space time geometry) with implications. For example, a black hole the size of an electron (

$m_e = 9.11 \times 10^{-31} kg$), imposes an acceleration of $a_{max} = 2.732 \times 10^{30} m/s^2$. So this accelerated frame would detect a Unruh radiation at 1.1×10^{10} K. Also the minimal length result implies the existence of the discreteness (granular nature) of space and cures the singularities that plague General relativity by imposing a general bound on length of $4.724 \times 10^{-36} m$.

In conclusion, a corrected Rindler space geometry directly proves an existence for the maximal acceleration and minimal length in quantum gravity, not only for the Schwarzschild metric with a horizon distance half of the Schwarzschild radius but also for the Reissner metric. Therefore the introduction of $\alpha^{1/2}$ in the formula for acceleration must be thoughtfully investigated as this solves all the problems brought about by the General relativity theory.

Experimental Search for Quantum Gravity

Derivation of the mass of an electron

- The size of Planck stars

(1)From the book Quantum Gravity in a Nutshell1 Chapter5 page89 (by Balungi Francis)

$$r_1 = \frac{4\pi\varepsilon\hbar^2}{me^2}n^2 = \frac{4\pi\varepsilon\hbar^2}{me^2}\left(\frac{\alpha_e}{\mu_e}\left(\frac{\omega^0_3}{8}\sqrt{3\pi}\right)^{1/2}\right)^2$$

Principle quantum number, $n = \frac{\alpha_e}{\mu_e}\left(\frac{\omega^0_3}{8}\sqrt{3\pi}\right)^{1/2}$, m is electron mass

(2) From the paper Carlo, R & F. Vidotto 2014-Planck Stars arXiv1401.6562

$$r_2 = \left(\frac{t_H}{3.48\pi t_p}\right)^{1/3} l_{p}$$

t_H –Age of universe and t_p –Planck time

- Electron mass

Fusing the two equations above, we have on Equating $r_1 = r_2$

$$m = \frac{4\pi\varepsilon\hbar^2}{e^2 l_p}\left(\frac{3.48\pi t_p}{t_H}\right)^{1/3}\left(\frac{\alpha_e}{\mu_e}\left(\frac{\omega^0_3}{8}\sqrt{3\pi}\right)^{1/2}\right)^2$$
$$= 9.11 \times 10^{-31} kg$$

The age of the universe is here $t_H \sim 71 billion\ years$

Chapter 2. Murder of Germans Sacred Cow: Experimental Test of General Relativity

It has long been suspected that the deflection of light in the vicinity of the sun exceeds the general relativistic predicted value of 1.75". An example of this, is the Erwin Finlay Freundlich 1929 solar eclipse expedition which produced a value of 2.24" larger than the general relativistic value. It is expected that once the reason for the deviation in the deflection angle has been found, it will disprove Einstein's imaginations for the curvature of space time. Although research into this field is scarce, we have managed through theoretical means under the assumption of a modified spherically symmetric solution to the Einstein field equation to prove E.F. Freundlich right. It is theorized that the bending of light near the sun is a function of the strength of the force (coupling constant) near the sun and the increasing distance from the sun's surface (in terms of the Schwarzschild radius).

It's almost hundred years since Sir Arthur Eddington experimentally proved Einstein's general relativity theory right. Since then, there has never been any competing theory that would prove Einstein wrong save for Loop quantum gravity and string theory. The fact that starlight is bent at the surface of the gravitating body by a deflection angle of 1.75" imposes a bound on the theoretical justification of gravity. Calculating an angle below or above 1.75" will be an upheaval in the founding blocks of physics. Erwin Finlay Freundlich was one of those people who stood out of the ordinary in 1929 when he published results with a larger angle of deflection than Eddington's.

An account on Freundlich 1929 expedition has been clearly given in Robert J.Trumpler and Klaus Hentschel papers as stated below;

"Among the various expeditions sent out to observe the total solar eclipse of May 9, 1929, that of the Potsdam Observatory (Einstein Stiftung) seems to be the only one which obtained photographs suitable for determining the light deflection in the Sun's gravitational field. Two instruments were used, but so far only the results of the larger one, a 28-foot horizontal camera combined with a coelostat, have been published. The three observers, Freundlich, von Klüber, and von Brunn, claim that these observations (four plates containing from seventeen to eighteen star images each) lead to a value of 2.24" for the deflection of a light ray grazing the Sun's edge; a figure that deviates considerably from the results of the 1922 eclipse, and which is in contradiction to Einstein's generalized theory of relativity".

The irreducible anomaly in the observations of the deflection of light by the sun has been known to exist since the birth of Einstein General relativity theory. For example, in a 1959 classical review by A.A.Mikhailov, it concludes that observations yield instead of a general relativistic prediction of 1.75arcsec at the limb of the sun the simple mean value of 2.03 ± 0.10 over the GR prediction (see the table below, from Mikhailov 1959 analysis)

Year	Number of plates	stars	Limits of r	Scale 1"=	p.e. of one star	A as given by observer	new reduction	A	r".75	straight line
1919	7	7	2·0- 5·4	18μ	0·15	1·98	2·07 ± 0·09	360	690	402
1921	4	71	2·1-13·0	22	0·13	1·72	1·43 ± 0·11	425	446	419
1929	2	18	1·5- 7·5	41	0·15	2·24	1·96 ± 0·08	1971	2372	3236
1936	2	29	2·0- 7·2	29	0·27	2·70	2·68 ± 0·37	1375	1534	1630
1947	1	51	3·3-10·2	30	0·24	2·01	2·20 ± 0·38	612	618	630
1952	2	11	2·1- 8·9	30	0·15	1·70	1·43 ± 0·18	7058	8693	4039

Weighted mean 1"·95 ± 0"·05 p.e.   Simple mean 2"·03 ± 0"·10

The existence of a 2.24" deflection angle by Freundlich, Von Kluber and Von Brunn therefore implies a requirement for the modification of the general theory of relativity. Science has evolved in this simpler manner of modifications although there are some who cling to the old thoughts of "The earth is the center of the universe and Einstein is always right". I am not proving anyone wrong but I want you to believe that the general relativity theory that was put forward by Einstein is not the only 'there is' excellent description of the universe, there are other ways far better than GR as it was with the Newtonian Gravitational force replacement with a curvature of space time.

K HENTSCHEL

FREUNDLICH in front (center) of the horizontal camera for measuring the deflection of light in the sun's gravitational field during the solar eclipse in Sumatra, 1929; from FREUNDLICH, V. KL/,)BER & V. BRUNN [1931]b PlateII or [1931]a p. 176.

In this section I will prove Erwin Finlay Freundlich solar eclipse results right but from a theoretical perspective. We base our study on the bending of starlight past the surface of the sun, we establish the deflection angle at which this occurs starting from General relativity and beyond.

Einstein's theory proposes that gravity is not an actual force, but is instead a geometric distortion of spacetime not predicted by ordinary Newtonian physics. The more mass you have to produce the gravity in a body the more distortion you get, this distortion changes the trajectories of objects moving through space, and even the paths of light rays, as they pass close-by the massive body. Even so, this effect is very feeble for an object as massive as our own sun, so it takes enormous care to even detect that it is occurring.

General Relativity predicts how much of this bending of light you should see given the mass of the object. By formula the Einstein General Relativity deflection angle is given by,

$$\theta_{GR} = \frac{4GM_\odot}{c^2 R_\odot} = 1.75 \text{arcsec}$$

Where, $M_\odot$ — mass of Sun(1.989×10^{30}kg)

$R_\odot$ — Radius of Sun(6.957×10^8m)

Knowing that the other fundamental forces are due to the exchange of specialized force-carrying particles: photons convey electromagnetism, the strong nuclear force is transmitted by gluons and the weak nuclear force is imparted by the movement of the W and Z bosons while gravity is due to the same kind of particle exchange-the graviton. Therefore we are forced to express Einstein's deflection angle in terms of the Planck length l_p and the gravitational coupling constant α_g as,

$$\theta_{GR} = 4\alpha_g^{1/2} \frac{l_p}{R_\odot} = 1.75 \text{arcsec} \tag{1}$$

Where, $l_p = 1.62 \times 10^{-35}$m and $\alpha_g = \frac{GM_\odot^2}{\hbar c}$

The gravitational coupling constant represents the strength of an interaction -the strength of the gravitational force, whereas the Planck length represents the shortest distance scale at which the graviton (if at all it exists) propagates in space. Therefore we have shown that, one can express the deflection angle in terms of the coupling constant. This idea is extracted from the previous chapter (1).

For a test particle or an observer falling freely from infinity to a distance r_o from the gravitating body, the modified spherically symmetric solution to the Einstein field equation will be given by;

$$(\theta R)^2 = \frac{1}{\alpha^{1/2}}\left(\frac{R_s r_o}{2}\right) \quad (2)$$

Where, $R_s = \frac{2GM}{c^2}$ is the Schwarzschild radius of a gravitating body, α is the coupling constant and θ is the angle of deflection of a light ray past a gravitating body.

This angle was never introduced in the previous chapter but it has been introduced in here for purposes supporting this research. Therefore the above equation reduces to Einstein general relativity prediction for the bending of light in a limit when $r_o = 8R_s$ and $\alpha = 1$ which gives the deflection angle as $\theta = \frac{2R_s}{R_\odot} = 4\alpha_g^{1/2}\frac{l_p}{R_\odot}$. This therefore means that, the effects of General relativity become apparent when the light ray is above the suns surface by the "Roche limit" in the photon sphere of only 23596.98m from the suns surface and under the influence of the strong nuclear force.

The Freundlich deflection angle might have taken a different twist than with Eddington 1.75arcsec result, which we are yet to find out and which is the reason for this expedition. Considering Eqn2 and noting parameters in a limit $r_o = R_s$ and $\alpha = \frac{K_e e^2}{hc} \sim \frac{1}{137}$, we deduce the deflection angle given by the formula,

$$\theta_Q = \frac{1}{2^{1/2}\alpha_e^{1/4}}\frac{R_s}{R_\odot} = \frac{2^{1/2}GM_\odot}{c^2\alpha_e^{1/4}R_\odot} \quad (3)$$

Substituting in the values of the constants, we get the deflection angle of a value,

$$\theta_Q = 1.025675 \times 10^{-05}\text{rad} = 2.12\text{arcsec}$$

This value is close to the Freundlich 1929 solar eclipse result and such is not a mistake but is a direct computation or deduction from a modified general theory of relativity on a sounding assumption that in a limit when the light ray past the sun is 2949.62m from the sun surface under the influence of an electromagnetic force in exchange of photons or bosons at $\alpha = \dfrac{1}{137}$ light will be deflected by an amount of 2.12" from its original course an amount larger than Einstein's value of 1.75".

In terms of the Planck length and the gravitational coupling constant Eqn(3) reads, $\theta_Q = \dfrac{2^{1/2}\alpha_g^{1/2}\, l_p}{\alpha_e^{1/4}\, R_\odot}$. But a clearer picture of this formula can be seen when we propose the existence of a length scale that depends of the mass of the body as,

$$\theta_Q = \alpha_g^{1/4}\frac{l_B}{R_\odot}$$

Where $l_B = \left(\dfrac{Mh}{ec^3}\right)^{1/2}\left(\dfrac{4G^3}{K_e}\right)^{1/4} = 5.295 \times 10^{-31}M^{1/2}$

This is a generalized length scale for all particles in the universe for example, for a Planck mass of $M_p = 2.178 \times 10^{-8}\text{kg}$ we have $l_B = 7.814 \times 10^{-35}\text{m}$, while for the sun of $M_\odot = 1.99 \times 10^{30}\text{kg}$, $l_B = 7.47 \times 10^{-16}\text{m}$ and lastly for a proton of $M_{pro} = 1.672 \times 10^{-27}\text{kg}$, $l_B = 2.165 \times 10^{-44}$.

In conclusion, the Einstein deflection angle of 1.75" is due to the presence of a strong nuclear force ($\alpha = 1$) at the surface of the sun in a range (radial distance) of $r_o = 8R_s$ from the suns surface where a light ray from a star is most close to the sun in its inner most stable orbit. Therefore the gluons act as a medium through which light from a star passes, hence proving Einstein wrong. The deflection of light is caused by how light from a star interacts with force carrier particles close to the surface of a gravitating body and on how distance increases from the sun.

The reason why the Freundlich 1929 deflection angle of 2.24" is larger than that of GR of 1.75" is because, at a radial distance of $r_o = R_s$ the ligth ray from a star interacts with the electromagnetic force carrier particles at the surface of the sun creating such a larger deflection. Therefore light from stars passes through a sea of gluons, bosons etc which dictate its course and this is a quantum effect that couldn't be explained by General relativity. The Freundlich value has been proved experimentally and there is no need for more expeditions except for the acceptance of a modified General relativistic theory that has been given above. To elaborate more, in the presence of the gravitational force at the surface of the sun we get a very small deflection angle (almost negligible), for example in a limit $r_o = R_s$ and $\alpha = \dfrac{GM_\odot{}^2}{\hbar c}$ we obtain a deflection angle given by, $\theta_R = 2^{1/2}\alpha_g{}^{1/4}\dfrac{l_p}{R_\odot} = 6.5 \times 10^{-20}{}''$. With the above analysis it is clear that the notion that the curvature of space near the Sun's surface bends or dictates the movement of light from distant stars is wrong and must be replaced with the newly given approach or a more modified theory.

Chapter3. How to Calculate a Mysterious Repulsive Force Pulling Galaxies Apart

It is known that in a homogenous cosmological universe, a positive cosmological constant induces repulsive forces. The question; is there a classical formula of the force of the cosmological constant like that of the gravitational force? How does the repulsive force relate to the cosmological constant and the coupling constant? How does understanding the energy density in relation to force, change the way we perceive Einstein's field equation? The section sets out to answer these and more questions about the cosmological constant problem. (see chapter 26 on the origin of gravity and electrodynamics)

Dr Lee Smolin represents the perimeter institute for theoretical physics. He claims that the mathematisation of physics has resulted in the reduction of the cosmos to a mathematical entity, which has not only confused physicists but accounts for their worst and most distracting assertions.

There is a wide spread speculation that the mathematical formulation of physics has not only confused physicists but has also lead to failures in the development of a quantum theory of gravity.

Although both general relativity and quantum mechanics work well in the domain of their applicability, it's unfortunate that there is no unified theory of gravity with quantum mechanics.

It is proposed that the unification of gravity with quantum mechanics will require us to change the kind of mathematics that was used by either Einstein or Schrödinger etal.. in the development of both theories. But why do we bother at all if there is another way in which we can express the theory better without the use of tensor fields.

The problem with the mathematical formulation of general relativity if at all it exists stems from the non existence of its experimental observation which wasn't the case with quantum mechanics. The formulation of quantum mechanics was based on the existence of experimental observations. Therefore quantum mechanics was founded on the existence of experiments which wasn't the case with general relativity. Einstein had to base his theorization on thought experiments which could or wouldn't be nearer to any experimental confirmation of the phenomenon being studied.

The same is also true for the formulation of quantum gravity. There is no sound experimental proof for the existence of quantum gravitational effects and therefore scientists like Hawking Stephen have also clung to the old formulations that were used by Einstein and his contemporaries to develop a quantum theory of gravity.

In this brief notice we show that an existence of a unified theory is rooted deep into the unnoticed pressure-energy density similar to the stress energy tensor appearing in Einstein field theory. Our major aim therefore is to provide proof for the questions set out below;

(i) If the cosmological constant introduces a force of repulsion between bodies. Is it true that the force increases in simple proportion to the cosmological constant and the coupling constant?

(ii) Is there a classical formula of the force of the cosmological constant like that of the gravitational force?

Einstein's general relativity equations famously described the curvature of space-time as the mechanism for gravity. In the original theory, Einstein added a "cosmological constant" that acted as an expulsive force to counteract gravity. That stabilized the universe so it didn't collapse in on itself, but Einstein abandoned the idea when further astronomical observations showed the universe was accelerating and not static, as the great physicist had thought.

Analogous to the known Einstein field equation, the curvature of space ?(cosmological constant) is here related to the energy density ω as,

$$\Lambda = \kappa\omega = \kappa\left(\frac{F^2}{8\pi\alpha\hbar c}\right) \quad (1)$$

Where $\kappa = \frac{8\pi G}{c^4}$ a constant appearing in Einstein's field equation, F is is the force in an interaction and α is the coupling constant.

The above expression implies that the cosmological constant is related to the force and therefore increases as a square of the force.

For the energy density in electric field, where $F = Ee$ and $\alpha = \frac{e^2}{4\pi\hbar c}$, the energy density will be given by, $\omega = \frac{F^2}{8\pi m\hbar c} = \frac{\varepsilon E^2}{2}$.

While for the energy density in the gravitational field, where $F = mg$ and $\alpha = \frac{Gm^2}{\hbar c}$, the energy density will be given by, $\omega = \frac{F^2}{8\pi n \hbar c} = \frac{g^2}{8\pi G}$. This can be written in simple terms as $\omega = \frac{\eta g^2}{2}$ where $\eta = \frac{1}{4\pi G}$.

From (1) therefore, the force responsible for the expansion of the universe is related to the cosmological constant by,

$$F = E_{pl}(\alpha\Lambda)^{1/2} \quad (2)$$

Where $E_{pl} = 1.9605 \times 10^9 J$ is the Planck energy.

Given the Planck (2015) values of $\Omega_\Lambda = 0.6911 \pm 0.0062$ and $H_0 = 67.74 \pm 0.46$ (km/s)/Mpc $= (2.195 \pm 0.015) \times 10^{-18}$ s^{-1}, Λ has the value of $1.11 \times 10^{-52} m^{-2}$ as given in wikimedia commons.

Based on the above given value, the force will then have a value of

$$F_{ob} = 2.0655 \times 10^{-17}(\alpha)^{1/2}$$

$$F_{ob} \sim 1.8 \times 10^{-18} N$$

This therefore is a force responsible for the expansion of the Universe. It is such a small force that will require sophiscated machines to measure. While the above force value is based on the fine structure constant, there is a value that is even smaller than that value by, $F_{ob} \sim 1.58 \times 10^{-36} N$ at $\alpha = 5.87 \times 10^{-39}$ between two protons.

However in quantum electrodynamics (QED) we compute a much larger value of $F_{QED} \sim 2.82 \times 10^{44}(\alpha)^{1/2} N$. This huge discrepancy is known as the cosmological constant problem. Therefore the relative strength of the force will be given by;

$$\frac{F_{QED}}{F_{ob}} \sim 10^{61}$$

The above value is in agreement with the Hubble age to the Planck time, which is the same as the total mass of the universe to the Planck mass as,

$$\frac{F_{QED}}{F_{ob}} = \frac{t_H}{t_{pl}} = \frac{M_U}{M_{pl}} \sim 10^{61}$$

The above given relationship implies a persistence constant error that is evident when comparing observational and theoretical calculations. This error needs to be distributed uniformly in order to correct for large discrepancies which accrue to calculated values in relation to observed values.

The problem lies in knowing the observed force value to the calculated value, since the force ratio doesn't correspond to the other ratios of time and mass. In other words changing the ratio $\frac{F_{QED}}{F_{ob}}$ to $\frac{F_{ob}}{F_{QED}}$ will cause other ratios to change.

It is therefore observed that the ratio of Hubble age to the Planck time and the total mass of the universe to the Planck mass will only be in line or tally with the Planck force to the Hubble force by a value $\sim 10^{61}$ and not otherwise.

Keeping other factors constant it is clear from the above given observations that the mysterious, repulsive force pulling galaxies apart is proportional to the coupling constant value in a given interaction. This proposal will be of such a great importance to the work of researchers involved in the field of quantum gravity

Chapter4: Is There A Limit To How Small Black Holes Can Become?

The smallest black hole would be one where the Schwarzschild radius equals the radius of a mass with a reduced Compton wavelength which is the smallest size to which a given mass can be localized. For a small mass M, the Compton wavelength exceeds half the Schwarzschild radius, and no black hole description exists. This smallest mass for a black hole is thus approximately the Planck mass, the micro black hole.

Contrary to the above observation, torsion (see Einstein-Cartan theory) modifies the Dirac equation in the presence of the gravitational field causing fermions to be spatially extended. This spatial extension of fermions limits the minimum mass of a black hole to be on the order of 10^{16}Kg, showing that micro black holes (of Planck mass) may not exist. Another mass limit is from the data of the Fermi Gamma-ray space telescope satellite which states that, less than one percent of dark matter could be made of primordial black holes with masses up to 10^{13}Kg.

The major aim of this section is to prove theoretically the existence of a minimum mass limit of a black hole and thereafter prove Chandrasekhar wrong (see Chandrasekhar 1983 Noble lecture concluding statement below)

"We conclude that there can be no surprises in the evolution of stars of mass less than 0.43Solarmass ($\mu_e = 2$). The end stage in the evolution of such stars can only be that of the white dwarfs. (**Parenthetically, we may note here that the so-called 'mini' black-holes of mass 10^{12} Kg cannot naturally be formed in the present astronomical universe.**)"

From the theory of white dwarf stars, the radius limit of a white dwarf of mass M is given by the following equation,

$$R_w = \frac{(9\pi)^{2/3} h^2}{8} \frac{1}{m_e G(m_{pro})^{5/3} M^{1/3}} \text{ (i)}$$

Where m_{pro} and m_e is the proton and electron mass respectively

In the same way the Planck mass is deduced (i.e by equating the Schwarzschild radius to the Compton wavelength) is the same way in which we are to prove the existence of the mass limit of a black hole. From **Chapter 6** we have a formula for the gravitational binding energy of a star given as,

$$E_g = \frac{2M_{pl}{}^3 m_e c^2}{M m_{pro}{}^2}$$, when this is equal to the Newtonian gravitational potential

energy $\frac{GM^2}{R}$ we obtain the radius which is the smallest size to which a given mass can be localized as,

$$R = \frac{G m_{pro}{}^2}{2 m_e c^2} \left(M / M_{pl}\right)^3 \quad \text{(ii)}$$

M_{pl} is the Planck mass

This can be rewritten in the form,

$$R = R_k \left(M / M_{pl}\right)^3$$, where $R_k = 1.135 \times 10^{-51} m$ which is smaller than

the Planck length of $1.62 \times 10^{-35} m$

Therefore equating Equation (i) to Equation (ii) we deduce the mass limit of a black hole as,

$$M = \left((9\pi)^2 M_{pl}{}^{21} / 64 M_{pro}{}^{11}\right)^{1/10} = 2.992 \times 10^{13} Kg$$

The value is in excellent agreement with other theoretical and experimental observations

The radius of this black hole from Equation (ii) is thus $2.934 \times 10^{12} m$ larger than the radius of the sun of $7 \times 10^8 m$.

In conclusion therefore the end stage in the evolution of a star can only be that of the black hole with a mass $2.992 \times 10^{13} Kg$ and size of $2.934 \times 10^{12} m$ in contrast with the Chandrasekhar observations.

Note that the radius given by Equation (ii), $R = \frac{G m_{pro}{}^2}{2 m_e c^2} \left(M / M_{pl}\right)^3$ above is similar to the Equation for the size of the Planck star that was given by Rovelli and Vidotto, $r = l_p \left(\frac{M}{M_{pl}}\right)^n$ where l_p is the Planck length and n is the positive number. This is a clear indication that there is a length that is smaller than the Planck length.

Chapter5. Particle Emission from Black Holes

Can we ever hope to find the right way? Nay more, has this right way an existence outside our illusions? I will answer without hesitation that there is, in my opinion, a right way, and that we are capable of finding it. Our experience hitherto justifies us in believing that nature is the realization of the simplest conceivable mathematical ideas. I am convinced that we can discover by means of purely mathematical constructions the concepts and the laws connecting them with each other, which furnish the key to the understanding of natural phenomena.

Albert Einstein (1879-1955)

In 1975 Hawking calculated quantum mechanically that a black hole will emit particles as if it were a black body with a temperature proportional to its surface gravity. Although this thermal emission is insignificant for black holes formed by stellar collapse, it is of crucial importance for the small primordial black holes formed by density fluctuations in the early universe.

The most significant consequence of a black hole is that, the temperature of a black hole increases as a black hole loses mass. The temperature increases exponentially into a burst of gamma rays leaving a black hole remnant. There is no clear account on this, not until we have fully developed a consistent quantum theory of gravity (where the mass of a black hole approaches the Planck scale of mass and radius). The evaporation of a black hole starts with a spin down phase in which the Hawking radiation carries away the angular momentum, after which it proceeds with emission of thermally distributed quanta until the black hole reaches the Planck mass.

The radiation spectrum contains all Standard Model particles, which are emitted on our brane, as well as gravitons, which are also emitted into the extra dimensions. It is expected that most of the initial energy is emitted during this phase in Standard Model particles.

One of the major problem with black holes is that, we cannot directly measure any properties of them neither can we produce black holes in any terrestrial experiment. According to Cheung (2002), this is due to the fact that in order to produce black holes in collider experiments one needs a centre of mass energy above the Planck scale, which is obviously inaccessible at the moment. But thanks to the introduction of the numerical coefficient, we can now as this book directs, detect a vast number of black holes in our galaxy by observing and detecting the mass scale or the low energy scale quanta emitted whenever a black hole evaporates due to stellar collapse.

The numerical coefficient α depends on which particle species can be emitted at a significant rate and can be determined by taking the effect of the absorption cross section. This coefficient is of great importance in the standard model and if described in detail it could unlock the secrets hidden deep in the cosmos.

The dominant contribution to α in the standard model comes from fermions, the contribution to α for electrons and positrons is 1.575×10^{-4} (Don Page 1975). Page calculated the emission rates for massless particles, predicted the lifetime of black holes (from the total power emitted in all modes) and also deduced the numerical coefficient for the dimensionally determined quantities (in terms of the Planck mass etc). The coefficient appears in Eqn25 and Eqn26 of the rate of change of mass and the life time of a black hole by Don Page (1975). There is no known formula relating the numerical coefficient to the mass scale or low energy scale quanta emitted the mass of an electron and the mass of a proton. Yet this could provide a unique probe of at least four areas of physics: the early Universe; gravitational collapse; high energy physics; and quantum gravity.

Assuming that a black hole emits particles at a mass scale M_* (low energy scale quanta), we propose the Numerical coefficient to be

$$\alpha = \frac{2M_* m_e}{m_p^2}$$

Where, m_p is the mass of the Proton and m_e is the mass of an electron. In the table below, we give values of α for different mass scales M_*.

Table 1

A   Mass scale M_* (kg) Remarks

2.837×10^{15} 4.343×10^{-9} Planck particle-Planck scale

1   1.531×10^{-24} Yet to be found $\approx 1TeV$

0.073 2.23×10^{-25} Higgs boson

1.575×10^{-4} 2.41×10^{-28} Pion-Neutral- cosmic rays

Note: M_{pl} (the Planck mass of $\left(\frac{\hbar c}{8\pi G}\right)^{1/2} = 4.343 \times 10^{-9} kg$)

From the analysis given above, a black hole of mass M_{BH} will have a temperature and a life time given by

Temperature:
$$T = \frac{M_{pl}^2 c^2}{kM_{BH}}\left(\frac{2M_* . m_e}{m_p^2}\right) = \frac{M_{pl}^2 c^2}{kM_{BH}}\alpha$$

$$T = 8.0305 \times 10^{46}\frac{M_*}{M_{BH}}$$

Lifetime:
$$\tau = \frac{GM_{BH}^3 m_p^2}{M_{pl}^2 c^3 M_* . m_e} = \frac{2GM_{BH}^3}{M_{pl}^2 c^3 \alpha}$$

$$\tau = 8.019 \times 10^{-43}\frac{M_{BH}^3}{M_*}$$

Note: The power of a black hole is given by, $P = \frac{M_{pl}^2 c^6}{2GM_{BH}}\alpha$

For purposes of this study, let us limit ourselves to two Black holes -Primordial, one with a mass of $4.7 \times 10^{11} kg$ and another with mass $1.331 \times 10^{17} kg$. We calculate the Temperature and life time of these black holes at known and assumed mass scales as given in table 2 and table 3.

Table 2.

Black hole (Kg)	Mass scale M_*	Temp-T(K)	$\tau_{(sec)}$	Remarks
	4.343×10^{-9}	7.42×10^{26}	19.17	Early universe
4.7×10^{11}	1.531×10^{-24}	2.62×10^{11}	5.44×10^{16}	Current Age of the Universe
	2.23×10^{-25}	3.81×10^{10}	3.73×10^{17}	Current Age of the Universe
	2.41×10^{-28}	4.12×10^{7}	3.46×10^{20}	

Note: For $M_* = M_{BH}$ we obtain the temperature of a Black hole $T = 8.0305 \times 10^{46} K$. This is the maximum temperature of a black hole above which the black hole cease to exist.

Table 3.

Black hole (Kg)	Mass scale M_*	Temp-T(K)	$\tau_{(sec)}$	Remarks
	4.343×10^{-9}	2.62×10^{21}	4.34×10^{17}	Current Age of the Universe
1.33×10^{17}	1.53×10^{-24}	9.24×10^{5}	1.23×10^{33}	
	2.23×10^{-25}	1.35×10^{5}	8.46×10^{33}	
	2.41×10^{-28}	145.52	7.83×10^{36}	

We learn from the above tables that, the temperature and lifetime associated with a black hole will not only depend on the mass of a black hole but also on the mass scale of the quanta emitted as scaled from the numerical coefficient which depends on which particle species can be emitted at a significant rate. For example, the theory of black hole radiations that was developed by S.W. Hawking will only become correct and deducible to the Hawking temperature and life time formula for black holes in a limit $M_* = 1.531 \times 10^{-24} kg$ and $\alpha = 1$. Such that, $T = \dfrac{1.229 \times 10^{23} kg}{M_{BH}}\ ^{\circ}K$ and $t = 5.238 \times 10^{-19} M_{BH}^{3}$. In other words M_* is assumed to be the scale of the underlying theory. The predictions of the Hawking radiations for a black hole with mass $4.7 \times 10^{11} kg$ are as given in table 2 at a mass scale

$1.531 \times 10^{-24} \text{kg}$. These take on similar properties for the Higgs boson. Therefore observations at such a scale could shed more light on the detection of a $4.7 \times 10^{11} \text{kg}$ black hole.

If we observe at a scale of a Pion $M_* = 2.41 \times 10^{-28} \text{kg}$ at the current age of the universe (about $13.8 \times 10^9 \text{yrs}$) we should be able to detect a black hole with a mass of $6.698 \times 10^4 \text{kg}$ (Primordial Black hole).

The ideas presented above could provide a unique probe of at least four areas of physics: the early Universe; gravitational collapse; high energy physics; and quantum gravity. The first topic is relevant because studying primordial black hole formation and evaporation can impose important constraints on primordial inhomogeneities and cosmological phase transitions. The second topic relates to recent developments in the study of "critical phenomena" and the issue of whether primordial black holes are viable dark matter candidates. The third topic arises because primordial black hole evaporations could contribute to cosmic rays, whose energy distribution would then give significant information about the high energy physics involved in the final explosive phase of black hole evaporation. The fourth topic arises because it has been suggested that quantum gravity effects could appear at the TeV scale ($M_* = 1.531 \times 10^{-24} \text{kg}$) and this leads to the intriguing possibility that small black holes could be generated in accelerators experiments or cosmic ray events, with striking observational consequences (see B.J.Carr, 2005).

Lastly, the significance of α-the numerical coefficient can be seen in a broad sense when applied to the sun. If we take the sun to be a black hole with mass $1.99 \times 10^{30} \text{kg}$ and a temperature at its center of $T = 1.5 \times 10^7 \text{K}$, we obtain a mass scale of $M_* = 3.717 \times 10^{-10} \text{kg}$, which gives a life time of $\tau = 1.700 \times 10^{58} \text{sec}$, the time that will be taken by the sun to dissipate if the temperature given at its center was $1.5 \times 10^7 \text{K}$.

The Chandrasekhar Mass Limit

A region in the universe has a potential energy of self-gravitation,

$$E_g = \frac{M_{pl}^2 c^2}{M_{BH}} \left(\frac{2M_* m_e}{m_p^2} \right)$$

A star will collapse to a White dwarf when the above energy is in equilibrium with the energy due to the electron degeneracy pressure of an Hydrogen atom given as,

$$E_e = m_e c^2 \left(\frac{\mu_e^2}{6.144\pi^3} \right)$$

where $\mu_e = 2$ is the average molecular weight per electron, which depends upon the chemical composition of a star.

Then for, $E_g = E_e$

$$M = \frac{12.288\pi^3 M_{pl}^2 M_*}{\mu_e^2 \, m_p^2}$$

In a limit for $M_* = M_{pl}$ and $\alpha = 2.837 \times 10^{15}$, we obtain the Chandrasekhar mass limit for a white dwarf star as,

$$M = \frac{12.288\pi^3 M_{pl}^3}{\mu_e^2 \, m_p^2} = 1.4 M_{sun}$$

Note that such a result is only possible in the given limit but for a limit such as $M_* = 1.531 \times 10^{-24} \text{kg}$ we obtain, $M = 4.956 \times 10^{-16} M_{sun}$ which is the mass of the Primordial black hole, providing evidence for the Hawking limit for particle emission by black holes as described previously

Planck epoch

From an expression for the life time of a black hole, it is theorized that a Black hole has a mass $M_{BH} = kM_*^{1/3}$ where k is a constant. For k=1, we have a life time of $8.019 \times 10^{-43} \text{sec}$ almost the Planck time-the earliest period of time in the history of the universe).

The Bekenstein-Hawking area entropy law

From the Black hole temperature we can calculate the entropy of a black hole,

The total energy of a black hole with mass M and surface area A is given as,

$$E = \frac{Ac^5 M_{pl}^2 M_* m_e}{2\pi G \hbar m_p^2 M}$$

The change in entropy when a quantity of E is added to a black hole is,

$$S = \frac{E}{T}$$

Since the temperature is known (see above) on substituting we have

$$S = \frac{Ac^3 k}{4\pi G\hbar}$$

This is the Bekenstein-Hawking area entropy formula.

Chapter6. A Brief Account on the Lowest Possible Energy State of the Hydrogen Atom and the Chandrasekhar Mass Limit

In this section we prove an existence of the minimal principal quantum number which imposes a general bound on the energy level of the Hydrogen atom and the orbital radius of an electron. The results are derived from general laws not known by the entire scientific community. The section therefore provides a relationship between the micro and macro structures of the universe at a level when the atomic mass limit is in equal proportion to the Chandrasekhar mass limit.

I won't go into details of the literature of the Chandrasekhar mass limit as these have been repeatdly written and analysed in almost a million papers about the topic. But for a brief introduction into the derivation of the Chandrasekhar mass I refer the reader to Chandrasekhar 1983 Noble prize lecture (1). Almost every aspect of a white dwarf star has been studied but there is one thing which we do not know about white dwarfs in relation to the Hydrogen atom and this is encoded in Flower's original statement;

"The black-dwarf material is best likened to a single gigantic molecule in its lowest quantum state. On the Fermi-Dirac statistics, its high density can be achieved in one and only one way, in virtue of correspondingly great energy content. But this energy can no more be expended in radiation than the energy of a normal atom or molecule. The only difference between black-dwarf matter and a normal molecule is that the molecule can exist in a free state while the black-dwarf matter can only so exist under very high external pressure".

The question is, do we have an existing relationship between the mass limit of the Hydrogen atom and the White dwarf star? If the black-dwarf material is best likened to a single gigantic molecule in its lowest quantum state, what is the lowest possible energy state at which such a relationship exists?

Briefly let us propose in formula a model to support our argument; Firstly, let the potential energy of self-gravitation of a star be given by,

$$E_g = \frac{2M_{pl}{}^3 m_e c^2}{M_g m_{pro}{}^2} \ (1)$$

Where M_{pl} is the Planck mass $\left(\frac{\hbar c}{8\pi G}\right)^{1/2}$, m_{pro} is the Proton mass, m_e electron mass, M_S mass of star and c is the constant speed of light

Secondly, let the energy due to the electron degeneracy pressure of an Hydrogen atom be given as,

$$E_e = m_e c^2 \left(\frac{\mu_e^2}{6.144\pi^3}\right) \quad (2)$$

Where $\mu_e = 2$ is the average molecular weight per electron, which depends upon the chemical composition of a star.

Lastly, the quantized energy of an Hydrogen atom is given by,

$$E_n = \frac{m_e K_e^2 e^4}{2n^2 \hbar^2} \quad (3)$$

Where n is the principle quantum number which indicates the energy levels in the Hydrogen atom.

By connecting the above three equations we shall be able to deduce the Chandrasekhar mass limit and the lowest principle quantum number in the Hydrogen atom, providing one of the first relationship between the microscope and macroscopic structures of the universe.

i)The Chandrasekhar Mass limit

Equating (1) to (2), when the gravitational potential energy of a star is in equilibrium with the energy due to the electron degeneracy pressure of a star, (i.e the Chandrasekhar limit is the mass above which electron degeneracy pressure in the star's core is insufficient to balance the star's own gravitational self-attraction).

$$\frac{2M_{pl}^3 m_e c^2}{M_S m_{pro}^2} = m_e c^2 \left(\frac{\mu_e^2}{6.144\pi^3}\right)$$

On cancelling like terms and arranging we have the Chandrasekhar mass limit as,

$$M_S = \frac{12.288\pi^3 M_{pl}^3}{\mu_e^2 \ m_{pro}^2} = 1.4 M_{sun}$$

ii)The lowest principle quantum number and energy level in the Hydrogen atom

Equating (2) to (3) we have

$$m_e c^2 \left(\frac{\mu_e^2}{6.144\pi^3} \right) = \frac{m_e K_e^2 e^4}{2n^2 \hbar^2}$$

On arranging and canceling like terms we have;

$$n = 1.753 \pi^{3/2} \left(\frac{\alpha_e}{\mu_e} \right) = 0.0356$$

Where $\alpha_e = \frac{K_e e^2}{hc} = \frac{1}{137}$ the fine is is structure constant

This is the allowed principal quantum number or the lowest energy state of an Hydrogen atom for a white dwarf star at the Chandrasekhar mass limit.

Therefore the quantized energy of the Hydrogen atom at this principal number is

$$E_n = \frac{13.606 eV}{n^2} = 10.74 \times 10^3 eV$$

And the electron radius at this energy level is $r = 6.69 \times 10^{-14} m$.

This result implies that, whereas the Bohr's orbital quantization doesn't permit orbits below the Bohr radius of $5.28 \times 10^{-11} m$, the theory above says that this is possible for an atom under high pressure. The electrons are therefore bound to the surface of the proton.

In another case where the Chandrasekhar mass limit is given by,

$$M_C = \frac{\omega^0{}_3 \sqrt{3\pi}}{2} \left(\frac{hc}{8\pi G} \right)^{3/2} \frac{1}{\mu_e^2 M_p^2}$$

Where $\omega^0{}_3 = 2.018236$, is a constant connected with the solution to the lane-Emden equation

The principal quantum number is given as

$$n = \frac{\alpha_e}{\mu_e} \left(\frac{\omega^0{}_3}{8} \sqrt{3\pi} \right)^{1/2}$$

$$n = 3.212 \times 10^{-3}$$

iii) **The relationship between the Hydrogen atom and white Dwarf-The atomic mass Limit**

Equating (1) to (3), when the gravitational potential energy of a star is in equilibrium with the total energy of an Hydrogen atom, we have a mass limit given as,

$$\frac{2M_{pl}^{3}m_{e}c^{2}}{M_{g}m_{pro}^{2}} = \frac{m_{e}K_{e}^{2}e^{4}}{2n^{2}\hbar^{2}}$$

$$M_{S} = \frac{4n^{2}M_{pl}^{3}}{\alpha_{e}^{2}m_{p}^{2}} \quad (4)$$

The result deduced above is similar to the Chandrasekhar mass limit value. It is therefore simple from the above result to impose a general bound on the principle quantum number, the energy level and the electron radius of orbit only if the Chandrasekhar mass limit value is equal to the value given in (4) above. This has been clearly deduced and the values are clear, that is, $n = 0.0356$, $E_{n} = 10.74 \times 10^{3}e$ and $r = 6.69 \times 10^{-14}m$. These values will always remain new and unique to the entire scientific community and will stand to prove the validity of the universal laws lying deep in equations 1-3. This paper is therefore proof that the microscopic and macroscopic cannot be separated in any physical theory.

Chapter 7. A Brief Account on the Implications of Quantum Gravity

The existence of a maximal acceleration in quantum gravity has great implications for physicists venturing into the field of quantum gravity. Although it has been studied in the previous chapters we again bring it here using different methods for purposes of imposing general bounds on the acceleration, length, mass, energy and temperature. The study of maximal acceleration is wide and that is why we are going to spend a great deal of time here in analyzing the consequences of its existence. The problem is important in calculating various limiting cases for both the theory of quantum gravity (at the Planck epoch) and the theory of quantum electrodynamics. As we shall see, all the calculations undertaken in the process lead us to one thing, a consistent quantum theory of gravity.

To differ from Newton's laws of motion and Einstein's theory of general relativity, our acceleration will depend on the dimensionless coupling constant which determines the strength of the force in any given interaction as was described earlier

$$a_{acel} = \frac{c^4}{e}\left(\frac{4\pi\varepsilon_0\alpha}{G}\right)^{1/2}$$

Where, c is the constant speed of light, e is the charge on an electron, α is the dimensionless coupling constant, ε is the permittivity of free space and G is the universal gravitational constant.

Various examples have been given below in which the theories of quantum gravity and quantum electrodynamics will act as limiting cases,

Maximal acceleration

For example, where the quanta exchanged between two electrons is a photon in the case of the electromagnetic force we have the electromagnetic coupling constant or the fine structure constant as, $\alpha = \frac{e^2}{4\pi\varepsilon_0 \hbar c}$ which deduces the acceleration to, $a = \frac{c^{7/2}}{(\hbar G)^{1/2}}$. This is the allowed maximum acceleration for quantum gravitational effects at the Planck epoch. But for the case where the quanta exchanged between two electrons is a graviton for a gravitational force, we have the gravitational coupling constant as, $\alpha = \dfrac{Gm^2}{\hbar c}$ which gives the acceleration on a quantum electrodynamics scale as,

$$a = \frac{m}{e}\left(\frac{4\pi\varepsilon_0 c^7}{\hbar}\right)^{1/2}.$$

Minimal length

Then the minimal radius to which a gravitating body or an electron can collapse in a commoving frame can also be deduced as, If we equate Newton's law of universal gravitation to our newly developed force as,

$$\frac{Gm^2}{R^2} = \frac{mc^4}{e}\left(\frac{4\pi\varepsilon_0\alpha}{G}\right)^{1/2}$$

We obtain the area as,

$$R^2 = \frac{me}{c^4}\left(\frac{G^3}{4\pi\varepsilon_0\alpha}\right)^{1/2}$$

then for $\alpha = \dfrac{Gm^2}{\hbar c}$, as in the first example, we obtain the minimum radius for a charged particle for quantum gravitational effects as,

$$R_{min1} = (Ge)^{1/2}\left(\frac{\hbar}{4\pi\varepsilon_0 c^7}\right)^{1/4} =$$

$4.717444838 \times 10^{-36}$m, or $R_{min1} = 0.2923 l_p$, where l_p is the Planck length. Then the fine structure constant will be calculated as,

$$\alpha = \left(\frac{R_{min1}}{L_p}\right)^4 = \frac{1}{(3.42155)^4} = \frac{1}{137.054}$$

Hence solving one of the unsolved problems in physics. But for $\alpha = \dfrac{e^2}{4\pi\varepsilon_0\hbar c}$, we obtain the minimal radius due to torsion in the gravitational field as,

$$R_{min2} = (m)^{1/2}\left(\frac{\hbar G^3}{c^7}\right)^{1/4}.$$

Minimal mass

On another note, we could derive the mass formula only if we equate the force $F = \dfrac{\hbar c^5}{G^2 m^2}$ to our force formula $F = \dfrac{mc^4}{e}\left(\dfrac{4\pi\varepsilon_0\alpha}{G}\right)^{1/2}$, then the mass expression is deduced as

$$m = \frac{\hbar^{1/3}e^{1/3}c^{1/3}}{(4\pi\varepsilon_0)^{1/6}G^{1/2}\alpha^{1/6}}$$

This gives the Planck mass at $\alpha = \dfrac{e^2}{4\pi\varepsilon_0\hbar c}$, Also the mass that incorporates all the constants of nature when $\alpha = \dfrac{Gm^2}{\hbar c}$ is deduced as,

$$m = \left(\frac{(\hbar c)^3 e^2}{4\pi\varepsilon_0 G^4}\right)^{1/8} = 1.177535 \times 10^{-8} \text{kg}.$$

This could be the mass of the graviton.

Energy

Then from Einstein's proposal for the radiation of the gravitational energy,

$$W = ma_{acel}R = \frac{mc^4}{e}\left(\frac{4\pi\varepsilon_0\alpha}{G}\right)^{1/2}R$$

we have an expression for energy as, , Where R is the radius of orbit of an electron around the nucleus of an atom, for $R \sim \frac{\hbar}{mc}$, and $\alpha = \frac{e^2}{4\pi\varepsilon_0\hbar c}$ we obtain the maximum energy as, $W = \left(\frac{\hbar c^5}{G}\right)^{1/2}$. This is the Planck energy at the Planck epoch.

But for $R \sim \frac{\hbar}{mc}$ and $\alpha = \frac{Gm^2}{\hbar c}$ we obtain the energy possessed by an electron of mass m in the electromagnetic field as, $W = \frac{2m}{e}\left(\pi\varepsilon_0 c^5\hbar\right)^{1/2}$, Then at the Planck epoch when $m = \sqrt{\frac{\hbar c}{G}}$ (the Planck mass), the energy required to accelerate an electron in the gravitational field will be given by, $W = \frac{2\hbar c^3}{e}\left(\frac{\pi\varepsilon_0}{G}\right)^{1/2}$ .In the case where the energy of the quantized states of the hydrogen atom $\frac{me^4}{16\pi^2 n\hbar^2\varepsilon_0^2}$, is equated to the energy of $\frac{2m}{e}\left(\pi\varepsilon_0 c^5\hbar\right)^{1/2}$, we obtain a crucial relationship between the fine structure constant and the principal quantum number n, as $\alpha = \left(\frac{n}{\pi}\right)^{2/5}$. Then the smallest quantum number that will give the value of the fine structure constant will be given by, n=$1.429101876 \times 10^{-5}$. This is the lower limit for the quantum theory.

Then on the quantum gravitational scale, in which the Bohr's quantized energy is equated to our energy $\frac{2\hbar c^3}{e}\left(\frac{\pi\varepsilon_0}{G}\right)^{1/2}$, we obtain the relationship between the principal quantum number n, the gravitational coupling constant α_G and the fine structure or the electromagnetic constant α_E as, $n^2 = \alpha_G\alpha_E^5$ or $n = \alpha_G^{1/2}\alpha_E^{5/2} \sim 3.493 \times 10^{-25}$. This value implies an upper bound on the energy states for a combined theory of gravity and quantum mechanics between two protons.

Temperature

Last but not least, we could write a modified Unruh- Davis effect as,

$$T = \frac{\hbar c^3}{k_B}\left(\frac{\varepsilon_0 \alpha}{\pi G}\right)^{1/2}$$

, when it is equated to the Hawking temperature effect

$$T = \frac{\hbar c}{k_B}(\Lambda)^{1/2}$$

, where Λ is the curvature of space, we obtain the curvature as,

$$\Lambda = \frac{c^4 \varepsilon_0}{\pi e^2 G}\alpha$$

, then for $\alpha = \frac{e^2}{4\pi\varepsilon_0 \hbar c}$, we obtain the maximum curvature for quantum gravitational effects as,

$$\Lambda = \frac{c^3}{4\pi^2 G\hbar}$$

. But when $\alpha = \frac{Gm^2}{\hbar c}$, we obtain the curvature for quantum electrodynamics effects as,

$$\Lambda = \frac{m^2 c^3 \varepsilon_0}{\pi e^2 \hbar}$$

.

Chapter8. A Semi- Classical Theory of Gravito-Magnetism

Albert Einstein was one of the physicists who attempted to develop a classical unified field theory but in vain. Other mathematicians and physicists who have attempted like Einstein to develop a unified field theory include among the many Hermann Weyl, Theodor Kaluza and R. Bach, but due to the continual development of a quantum theory and the difficulties encountered in developing a quantum theory of gravity most of the physicists have gave up working on the unified field theories as of date.

Although most of the scientists have abandoned classical theories, still they remain the only means through which the quantum theory of gravitation can be created and thereafter be unified with the other fundamental theories in physics as this book directs.

At present, one of the deepest problems in theoretical physics lies in harmonizing the theory of general relativity, which describes gravitation, and applications to large-scale structures (stars, planets, galaxies), with quantum mechanics, which describes the other three fundamental forces acting on the atomic scale.

However there is a connection between the fundamental forces of electromagnetism and gravitation through which the quantum theory of gravity can be created. This connection is based on two rules, 1) the fundamental significance of the finite and invariant velocity of light in inertial reference frames in the special theory, and 2) the reliance of the general theory of relativity upon the special theory of relativity locally in space-time. The connection between the fundamental forces of electromagnetism and gravitation follows immediately from these two points (Douglas M. Snyder). To unite electromagnetism with gravity and thereafter with the quantum theory, we must create in principle a clear and analytical study of the intensity of the electromagnetic wave from scratch.

Gravitational radiation is produced when massive bodies accelerate. This radiation is difficult to detect due to the weakness of the gravitational force. It can only be detected under vigorous observations of the radiations from supernovae and collisions of black holes. The study of the gravitational radiation would come straight from the Bohr's theory of an atom but it proves difficult since one cannot even deduce the intensity of the electromagnetic wave from such a theory. Rather the intensity of radiation emitted from an atom is studied using the known formula for the intensity in electromagnetism (EB /μo). To clearly understand the intensity of the electromagnetic wave, one needs to develop a formula for the intensity of a wave on a quantum scale. Once such is formulated, it would then become easy to perform calculations for the intensity of the gravitational waves.

The unification of electromagnetism with gravity for a long time has been difficult, in a statement written by Jeroen Burgers (2009) it is clear that the two interactions can be unified into a single interaction. In his statement Jeroen writes, "If an electrical charge is accelerating, it will emit radiation, e.g, as radio waves from an antenna. Correspondingly, according to GR, accelerating masses should emit gravitational radiation, a propagating deformation of space-time".

To differ from Bohr's model of an hydrogen atom, it is hereby theorized, that an electron moving in an atom will possess an energy due to the electric, magnetic and gravitational forces acting on it as,

$$W_p = \frac{mgEe}{Bev}r = \frac{F_G F_e}{F_B}r \tag{1}$$

Where $F_G = mg$ is the gravitational force, $F_e = Ee$ is the electric force on an electron in vicinity of an electric field and $F_B = Bev$ is the magnetic force

To determine the strength of the electromagnetic force on a quantum scale, we borrow an analogy from the theory of quantum mechanics by which the quantized angular momentum is deduced from the fine structure constant and denoting the coupling constant to behave as the principle quantum number (Remember our aim here is to deduce the intensity of a wave emitted from an atom due to an electron performing Bohr's orbits), In formula we express the quantized angular momentum of an electron due to an electromagnetic interaction as,

$$\frac{Ke^2}{c} = n\hbar \tag{2}$$

Where ħ is the reduced plank constant, c is the speed of light, k is the coulomb constant (k=1 /4πεo, εo is permeability of free space) and n is the principle quantum number.

Since the gravitational force is almost negligible in an atom, it becomes a catastrophe to treat it as a quantum mechanical effect. In quantum mechanics the angular momentum of an electron is quantized in units of $n\hbar$ while in gravitational mechanics, there is no such thing as quantization, which is why we treat gravitation classically. Then the formula for the angular momentum due to the gravitational force will be given by,

$$\frac{Gm^2}{c} = mvr \quad (3)$$

It is therefore evident that the gravitational descriptions of an electron can only be treated classically. This is why it has proven difficult to merge gravity with quantum mechanics. However such a problem has been solved here, by considering the assumptions below,

The speed of light in both quantum and gravitational processes is a constant and therefore if we substitute the speed of light from eqn (2) into eqn(3) we get the expression for the angular momentum as,

$$mvr = \frac{F_G}{F_e} n\hbar \quad (4)$$

We have thus introduced the ratio of the gravitational force to the electric force in the formula for Bohr's quantized angular momentum. This ratio represents the negligible gravitational effects in an atom. It is therefore a correction to the Bohr's atomic model.

Because the gravitational force can be expressed in many ways by using Eqn(1), we can deduce the power carried by the electromagnetic wave due to the motion of an electron in an atom. Making F_G the subject from equation (1) and substituting for it in equation (4), we get the power as,

$$F_B c = \frac{2\pi r^2 \lambda mv F_e^2}{nh^2} \quad (5)$$

Since the de Brogile wave length is $\lambda = \frac{h}{mv}$, and the surface area of the sphere is $A = 4\pi r^2$. Then the Intensity of a wave from a particle exhibiting both wave and particle properties is

$$\frac{F_B c}{A} = \frac{F_e^2}{2nh} = \frac{E^2 e^2}{2nh} \quad (6)$$

Keeping other factors constant we have theorized that, the intensity of a wave is proportional to the square of the electric field, a fact that would be impossible to deduce in Bohr's atomic model. The above formula can only be deduced if only we take into account (in theory), the combined effects of gravity and electromagnetism.

On the other hand, If we let the power of the electromagnetic wave be P= FBc, and n be the fine structure constant $\alpha = ke^2/\hbar c$, then the equation for the intensity of the electromagnetic wave comes out clearly as, $P = EB /\mu_0 = 2\varepsilon_0 E^2 c$, Where μ_0 is the permeability of free space.

Gravitational waves are harder to generate than the electromagnetic waves, simply because, due to the conservation of the angular momentum, there's no dipole gravitational radiation meaning that, the dominant mode of the gravitational radiation is quadrupole radiation (see Einstein quadrupole formula). Although the gravitational waves are hard to generate, there is a possibility of studying their character and property by calculating their intensity, a fact that has proved to be difficult since there is no formula in quantum gravitational theories that can prove it.

The intensity of the gravitational wave can be deduced if only we assume that, the electric force in quantum mechanics can only be expressed as, $(Ee = \frac{n^2\hbar c}{8\pi r^2})$. Where r is the radius of orbit of an electron in an atom, when this radius approaches the schwarzichild's radius $(r = \frac{Gm}{c^2})$ for Black holes, then the electric force is expressed as, $(Ee = \frac{n^2\hbar c^5}{8\pi G^2 m^2})$. But the intensity in quantum mechanics was deduced in Eqn6. Therefore the intensity of the gravitational wave will be given by,

$$\frac{F_B c}{A} = \frac{E^2 e^2}{2nh} = \frac{n^3 \hbar c^{10}}{256\pi^3 G^4 m^4 7}$$

At the surface of the sun the intensity of the solar radiation is about $6.33 \times 10^7 W/m^2$, which means that, we will require a mass of about $8.8965 \times 10^{19} kg$, to obtain that intensity from our formula above.

For the sun of mass $1.989 \times 10^{30} kg$, the intensity is so small with a value of only, $0.253 \times 10^{-33} W/m^2$. This indicates that it will be difficult without sophiscated instruments for us to measure the effects of quantum gravity.

Chapter9. On The Quantum Electrodynamics And Quantum Gravity Magnetic Field Limits.

The scale in quantum electrodynamics (QED), above which the electromagnetic field is expected to become non linear, also called the Schwinger limit, was first derived by Fritz Sauter in 1931. However In this section we develop a mechanism (which differs from Fritz's approach) through which the Schwinger limit is deduced using a dimensionless number, which gives the critical magnetic field in quantum electrodynamics when its value is equal to the electromagnetic coupling constant and in the same way gives the critical magnetic field in Quantum gravity when its value is equal to the gravitational coupling constant.(see Applications in Chapter 25 and 26)

According to D.A. Leahy, the application of quantum electrodynamics in strong magnetic fields only fairly recently has been a subject of interest. The motivation for this study was the discovery of Neutron stars with very high magnetic fields of orders 10^{12} -10^{13}G.

With the discovery of magnetars, quantum electrodynamics calculations which are valid for very high fields became of great interest. The critical value of the magnetic field is defined as $B = \frac{m^2 c^3}{\hbar e} = 4.414 \times 10^{13} G$ .However, there is a value of the magnetic field that is bigger and stronger than the critical magnetic field strength in Quantum electrodynamics and this magnetic field is of orders of magnitude $10^{52} G$. Such a big value has not been deduced in any existing scientific literature and that is the reason why I take pleasure in deriving it here and hence call it the "quantum gravity threshold".

From equation (6) $\frac{F_B c}{A} = \frac{F_e^2}{2nh}$, if we let the magnetic force to be equal in magnitude and strength to the electric force, we create two relationships, 1) the force on a particle falls off as the area it occupies and 2) the force falls off as the principle quantum number.

$$\text{Force (F)} = \frac{2hc}{A} = \frac{Bev}{n}$$

If n was the fine structure constant ($ke^2/\hbar c$, $k = 1/4\pi\varepsilon_0$), the speed of light in vacuum being $c = \lambda f = \lambda\omega/2\pi$ and the velocity of a particle in the magnetic field is $v = \omega r$ where ω is the angular frequency for circular motion we have

$$\frac{F_c}{F} = \frac{em}{2B\lambda\hbar\varepsilon_0} \qquad 8$$

Where $F_c = m\omega^2 r$ is the centripetal force

We have thus derived a general formula for the coupling of forces. Then the Schwinger limit in quantum electrodynamics for the critical magnetic field can be deduced from the above expression when we set the ratio of the forces to be equal to the electromagnetic coupling or fine structure constant as, $\frac{F_c}{F} = \frac{ke^2}{\hbar c}$

$$B_{QED} = \frac{2\pi mc}{\lambda e}$$

For a particle with deBrogile wavelength $\frac{2\pi\hbar}{mc} = \lambda$, the quantum electrodynamics threshold is given by,

$$B_{QED} = \frac{m^2c^2}{\hbar e} = 4.3697 \times 10^{13}G \qquad 9$$

However, for $\frac{F_c}{F} = \frac{Gm^2}{\hbar c}$, the gravitational coupling constant, and $\frac{2\pi\hbar}{mc} = \lambda$, the deBrogile wavelength, the quantum gravity threshold is given by a value,

$$B_{QG} = \frac{ec^2}{4\pi G\hbar\varepsilon_0} = 1.8423 \times 10^{52}G \qquad 10$$

We have thus deduced the constant magnetic field carried by an electron in the combined quantum electromagnetic and gravitational fields. The fact that the formula has the fundamental constant of electricity (ε_0), relativistic quantum mechanics (c, $\hbar$) and Gravity (G), is an indication that this is the quantum gravity limit or a scale at which the electromagnetic field is expected to become non linear.

In conclusion, the dimensionless coupling constant is therefore the ratio of the universal constant magnetic field (B_0) on the particle of mass m to the Schwinger magnetic induction limit- the strong magnetic field B external to virtual electron-positron pair enclosed in a quantum vacuum. In general, the coupling constant can be written as,

$$\alpha = \frac{em}{2B\lambda\hbar\varepsilon_0}$$

For $\lambda = \frac{2\pi\hbar}{mc}$, the coupling constant reads as, $\alpha = \frac{em^2c}{4\pi B\lambda\hbar^2\varepsilon_0}$

Implying, $\alpha = \frac{B_0}{B}$

Where $B_0 = \left(\frac{m}{\hbar}\right)^2 \frac{ec}{4\pi\varepsilon_0} = 3.22 \times 10^{11}G$,

This value of the magnetic field is smaller than the Schwinger limit and therefore has implications for the theory of QED. Once such a value is discovered in experiments it will lead to a complete observation of the electron-positron pairs. Finally the fine structure coupling constant and the gravitational coupling will manifest in situations where the magnetic field B is either a Schwinger limit given by equations 9 and 10 above. This implies that, the value of the magnetic field B_0 derived above is the same everywhere (i.e both in applications of electromagnetism and gravity)

CHAPTER10. On The Derivation of the Temperature and Entropy of Black Holes

If the semi-diameter of a sphere of the same density as the sun were to exceed that of the sun in the proportion of 500 to 1, a body falling from an infinite height towards it would have acquired at its surface greater velocity than that of light, and consequently supposing light to be attracted by the same force in proportion to its vis inertiae, with other bodies, all light emitted from such a body would be made to return towards it by its own proper gravity (John Michell).

The development of general relativity followed a publication of acceleration under special relativity in 1907 by Albert Einstein. In his article, he argued that any mass will "Distort" the region of space around it so that all freely moving objects will follow the same curved paths curving toward the mass producing the distortions. Then in 1916, Schwarzschild found a solution to the Einstein field equations, laying the groundwork for the description of gravitational collapse and, eventually, black holes.

By definition, a black hole is an astronomical object with a very strong gravitational effect, which disturbs particles across its event horizon. It is also true from the theory of general relativity, that even light cannot escape its gravitational pull. These objects have puzzled the minds of great thinkers for many years. History puts it that, they were first predicated by John Michell and Pierre-Simon Laplace in the 18th century but David Finkelstein was the first person to publish a promising explanation of them in 1958 based on Karl Schwarz child's formulations of a solution to general relativity that characterized black holes in 1916.

In 1971, Hawking developed what is known as the second law of black hole mechanics in which the total area of the event horizons of any collection of classical black holes can never decrease, even if they collide and merge. This is similar to the second law of thermodynamics which states that, the entropy of a system can never decrease. In 1972 Bekenstein proposed an analogy between black hole physics and thermodynamics in which he derived a relation between the entropy of black hole entropy and the area of its event horizon.

In 1974, Hawking predicted an entirely astonishing phenomenon about black holes, in which he ascertained with accuracy that black holes do radiate or emit particles in a perfect black body spectrum. Hawking was able to produce in result the temperature of a black hole and proposed that, this temperature was proportional to the surface gravity of a black hole.

Both the temperature and entropy of a black hole have been deduced in literature using different approaches which have proved to be true but only lack one ingredient and that is, the description of a force like the Newtonian force of gravity, which acts to pull matter from the event horizon up to the point of its singularity. If this force was true it would be able under general conditions to derive the Reissner- Nordstrom metric for the charged non-rotating black hole as we are to see.

Temperature of a black hole

It is here by hypothesized that, the gravitational field will create particles and emit them only if the electromagnetic force of such particles were equal to the force (unknown in literature) $F = \frac{Me}{r}\sqrt{\frac{Gp}{2\hbar\varepsilon_0\lambda}}$ .Where p, is the momentum of a particle.

Then under general conditions, the force given will reduce to the Reissner-Nordstrom metric as given here, if the momentum of an electron at a distance r from the singularity point to the event horizon is related to the de Brogile wavelength as $p = \frac{2\pi\hbar}{\lambda}$, and both the distance r and wavelength λ was the product of the speed of light c and the period T as r=cT and $\lambda = cT$, then the force will be given by $F = \frac{Mp}{r\hbar}\sqrt{\frac{Ge^2}{4\pi\varepsilon_0}}$, but since $\frac{p}{2\pi\hbar} = \frac{1}{\lambda}$, then we have,

$F = \frac{2\pi M}{T^2}\sqrt{\frac{Ge^2}{4\pi\varepsilon_0 c^4}}$, this reduces to $F = \frac{2\pi M}{T^2}r_q$, where $r_q = \sqrt{\frac{Ge^2}{4\pi\varepsilon_0 c^4}}$ is the Reissner-Nordstrom radius of a charged black hole.

Having derived the Reissner-Nordstrom metric from our force formula, we now return to our exercise of deriving the temperature of a black hole. We consider a particle with charge e, exhibiting deBrogile wave properties of momentum and wavelength from the centre of mass M of a black hole. We then assume that this particle experiences an electromagnetic force due to the magnetic and electric field created by other particles in its surrounding area. The same particle also experiences a force due to the strong gravitational field emanating from the black hole. Equating the two forces as $\frac{Me}{r}\sqrt{\frac{Gp}{2\hbar\varepsilon_0\lambda}} = \frac{e^2}{4\pi\varepsilon_0 r^2}$, from this expression we obtain the momentum of a particle as $p = \frac{\hbar e^2 \lambda}{2\pi A \varepsilon_0 G M^2}$. This is the momentum possessed by a particle (emitted by the gravitational field of a black hole) at the surface of the event horizon, where $A = 4\pi r^2$ is the spherical surface area of the horizon.

For relativistic effects, the kinetic energy of a particle will be related to its momentum by K.E=pc and to the Boltzmann's law by K.E=kT, where k is the Boltzmann's constant and T is the absolute temperature. By similarity we can equate the two energies as pc=kT, then from the equation of momentum we can obtain the temperature as,

$$T = \frac{\hbar e^2 \lambda c}{2\pi A \varepsilon_0 G M^2 k}.$$

Expressing the permittivity of free space in terms of the permeability of free space $\varepsilon_0 = \frac{1}{\mu_0 c^2}$, we obtain the Hawking temperature of a black hole as,

$$T = \left(\frac{4e^2 \mu_0 \lambda}{AM}\right)\frac{\hbar c^3}{8\pi G M k}$$

In a more general form, in terms of energies it can be expressed as,

$$T = \left(\frac{4e^2 \lambda}{A\varepsilon_0 M c^2}\right)\frac{\hbar c^3}{8\pi G M k} \; 11$$

Entropy of a black hole

In an attempt to prevent the violation of the generalized second law of thermodynamics, Bekenstein proposed a universal upper bound on the ratio entropy to energy for bounded systems (Phys RevD23, 287-1981), which was later rejected by Unruh and Wald in 1982. They proposed a thought experiment in which a box lowered down into a black hole felt an effective buoyancy force which was caused by the acceleration radiation felt by the box near the black hole. They argued further that, this buoyancy force would guarantee a lower bound on the energy gain of the black hole, hence saving the generalized second law without a need for entropy bound.

In this section we give a formula for the buoyancy force which is different from the Unruh and Wald formula which appeared in their 1982 paper. At a distance r from the center of mass m of a black hole, the buoyancy force is given by,

$$F_B = \frac{rc^6}{8G^2 m} \quad 12$$

From the above force formula the energy gain by the black hole will be given by,

$$W_B = \frac{Ac^6}{32\pi G^2 m}$$

Where, A is the area of the event horizon. Since entropy is the ratio of energy to temperature, $S_B = W_B / T_B$ and temperature of a black hole is known from equation 11, then the entropy of a black hole is given by,

$$S_B = \frac{Akc^3}{4\,G\hbar} \left(\frac{A\varepsilon_0 Mc^2}{4e^2 \lambda} \right) 13$$

The numerical coefficient $\frac{4e^2 \lambda}{A\varepsilon_0 Mc^2}$ appearing in the above formulas is related to the numerical coefficient that was given in the first chapter of this book. Its significance is left for researchers to find out.

Chapter 11. A Unique Theory of Black Hole Thermodynamics

A Black hole is a mathematically defined region of space time exhibiting such a strong gravitational pull that no particle or electromagnetic radiation can escape from it. Many theories have been created to explain the properties of the black hole but the theory created here is far more different from the other theories although it may give the same results. Using a quite different approach towards solving a problem is efficient since it comes with it new predictions in the process which could have been hidden in other approaches. Below we try to present adhoc proofs-laws that may be of help in building our theory about black holes, note; these proofs can be derived mathematically from equations 1 up to 4 above but for purposes of simplicity they have been listed here below, however their derivations will be given in the last chapters of this book.

The laws or equations:

It is well known that the electric field is force per unit charge but here a generalized equation for an electric field created by an electron exhibiting wave properties in the nucleus of an atom in the gravitational field on a quantum scale is given by

$$E = \frac{1}{r}\sqrt{\frac{Gm^3 f}{2\hbar\varepsilon_0}} \quad (g)$$

Then the electric force in this case will be formulated as

$$F_1 = \frac{e}{r}\sqrt{\frac{Gm^3 f}{2\hbar\varepsilon_0}} \quad (h)$$

The surface area at a radius r of orbit of an electron of mass m around the nucleus of an atom in a wave like manner is given by

$$\text{surface area}(A) = \frac{\lambda\mu_0 e^2}{m} \quad (i)$$

The time taken by the magnetic field B of an electron to pass through a given surface is

$$\text{time}(t) = \frac{\lambda\varepsilon_0 AB}{e} \quad (j)$$

Note: the above expression is the same as Faraday's induction law.

The gravitational force acting on all matter in the universe or the modified gravitational force is given as

$$F_2 = \left(\frac{Gm^3}{r^2}\right)\left(\frac{e}{2B\lambda\hbar\varepsilon_0}\right) \quad (k)$$

The above formulas are important in deriving the formula for the temperature, entropy and the time taken by a black hole to evaporate as shown below;

Temperature of a black hole

It is known that the kinetic energy KE of molecules in the Boltzmann hypothesis is related to the temperature of the body in question in this case a black hole (in relation to the black body) by $KE = \varphi T$ where φ is Boltzmann's constant. The formula for the kinetic energy can be derived by using a hypothesis that the electromagnetic force – coulombs force is equal to eqn(h) as

$$\frac{ke^2}{r^2} = \frac{e}{r}\sqrt{\frac{Gm^3 f}{2\hbar\varepsilon_0}}$$

On squaring both sides of the equation, cancelling like terms and taking into account that the frequency of an electron is $f = \frac{v}{\lambda}$, then the kinetic energy of an electron inside the black hole is given by

$$KE = \frac{\lambda\mu_0 e^2}{A} \frac{c^3\hbar}{8\pi Gm^2}$$

Since the surface area is given as from eqaution(i) then the kinetic energy of molecules or particles (for an ideal gas) within the black hole will be given by

$$KE = \frac{c^3\hbar}{8\pi Gm} = T\varphi \quad (l)$$

Then from Boltzmann's relationship the temperature of the black hole is formulated as

$$T = \frac{c^3\hbar}{8\pi Gm\varphi} \quad (m)$$

The entropy of the black hole

By definition entropy is a measure of disorder. To solve the entropy of black holes we shall consider a very complex argument about the entropy in question. We assume that the modified gravitational force given by equation (k) is identical to the modified electric field given by equation(h) as,

$$\left(\frac{Gm^5}{r^2}\right)\left(\frac{e}{2B\lambda\hbar\varepsilon_0}\right) \equiv \frac{e}{r}\sqrt{\frac{Gm^5 f}{2\hbar\varepsilon_0}}$$ in otherwise the two forces are equal but opposite. Then squaring both sides of the equation and multiplying through by Gc^5 one obtains a new relation of forces on both sides given as

$$\frac{tc^7}{16\pi G^2 m} = \frac{Ac^6}{32\pi rm\, G^2}$$

Both the left and right hand side represent a force. From the left hand side t is the expression of time given by $$t = \frac{\hbar e^2}{2m^5 c^2 G\varepsilon_0}.$$ Note: the left hand side force is the pull of matter inside the black hole while the right hand side force is the force acting on particles or matter at the surface of the black hole.

Since the heat is the product of the force on a particle and the distance r from the centre of the black hole, then using the force on the right hand side of the above equation the heat will be given by

$$Q = \frac{Ac^6}{32\pi m\, G^2}$$

Remember the temperature of the black hole is also known from equation (m) and by definition the entropy of the system is the change in heat per unit temperature $\frac{Q}{T}$, then the entropy of the black hole will be given by

$$S = \frac{A\varphi c^3}{4G\hbar} \text{ (n)}$$

This implies that the entropy of a black hole is proportional to its surface area.

The time taken by a black hole to evaporate

Assuming that particles that formed a black hole are moving away or are separating from it after a given time of its existence, if we measure the relative speed of these particles in relation to the energy they carry we obtain a relationship given by

$$\frac{v^2}{c^2} = \frac{8\pi G}{c^2}\left(\frac{W}{8\pi r}\right) \text{(o)}$$

Where v is the velocity of these particles as measured relative to the speed of light c and W is the energy carried by the particles as they move away from the centre of the black hole at a distance r.

If we let the force causing the particles to separate from the black hole be given as $\frac{Gm^3 e}{2r\lambda B\hbar\varepsilon_0}\frac{v}{c}$, then the energy of these particles will be given by

$$W = \frac{Gm^3 e}{2r\lambda B\hbar\varepsilon_0}\frac{v}{c}$$

Substituting this in equation (o), we obtain a relationship of time as given by the law 3 of equation (j) as

$$t = \frac{v^2}{c^2}\left(\frac{\pi G^2 m^3}{\hbar c^4}\right)$$

The velocity of the particles in the astronomical lab will be measured as v= 4.193E6 m/s and since the speed of light is a constant then the time taken by a black hole to evaporate is given by,

$$t = \frac{5120\pi G^2 m^3}{\hbar c^4}$$

Chapter 12. A Unified Bohr and Quantum Gravity Theory

In the early 20th century[1], Ernest Rutherford[2] experiments established that atoms[3] consisted of a diffuse cloud of negatively charged electrons[4] surrounding a small, dense, positively charged nucleus. Given his experimental data, it was quite natural for Rutherford to consider a planetary model for the atom, the Rutherford model[5] of 1911, with electrons orbiting a sun-like nucleus. This model was a difficulty. The laws of classical mechanics predict that the electron will release electromagnetic radiation[6] as it orbits a nucleus. Because the electron would be losing energy, it would gradually spiral inwards and collapse into the nucleus. This was a disaster, because it predicted that all matter was unstable.

Can the macroscopic realm of gravity ever be merged with the strange microscopic kingdom of quantum particles to create a Grand Unified Theory of Everything?

To overcome this difficulty, Niels Bohr[7] proposed, in 1913[8], what is now called the Bohr model of the H atom. The model's key success laid in explaining the Rydberg formula[9] for the spectral emission lines[10] of atomic hydrogen. Not only did the Bohr model explain the reason for the structure of the Rydberg formula, but it provided a justification for its empirical results in terms of fundamental physical constants.

This section looks at the model in a very different way than that of Bohr. The fact that all accelerated particles do emit electromagnetic radiations is

1. http://en.wikipedia.org/wiki/20th_century

2. http://en.wikipedia.org/wiki/Ernest_Rutherford

3. http://en.wikipedia.org/wiki/Atom

4. http://en.wikipedia.org/wiki/Electron

5. http://en.wikipedia.org/wiki/Rutherford_model

6. http://en.wikipedia.org/wiki/Electromagnetic_radiation

7. http://en.wikipedia.org/wiki/Niels_Bohr

8. http://en.wikipedia.org/wiki/1913

9. http://en.wikipedia.org/wiki/Rydberg_formula

10. http://en.wikipedia.org/wiki/Emission_line

taken into account and therefore the acceptance for the unstableness of all matter is considered in due respect. In fact Bohr's ideas never required classical mechanics simply because it could not conform to the experimental observations of the spectrum of the Hydrogen atom that were obtained by Rydberg using his formula.

To merge gravity with Planck's quantum theory by then was also a problem at hand and therefore Bohr had to forego the problem by introducing in his theory adhoc postulates, and this could have been the reason why Einstein found problems in merging gravity with electromagnetism in what is called "The Grand unified field theory", of which he had to question the problem with the quantum theory and therefore request for a complete quantum theory. From Bohr's model many theories have been formed each building from the ideas of the model, but a certain point is reached where the theories can not conform well to the known laws of nature and therefore regarded as failures, which of course in their judgments is true. The problem is seen to come from exactly the roots of quantum mechanics.

The aim of this section is therefore to produce a generalized theory of atomic structure that incorporates in it gravity and quantum mechanics and thus predict the properties of the universe at the Planck era.

The Hydrogen atom exists in certain stationary states of discrete energies. The acceleration due to gravity of an electron in orbit around the nucleus will cause the atom to emit radiations (radiate energy) and thus make the atom unstable. The acceleration (g) falls off with time t provided the radius of orbit of the electron R is a constant thus the acceleration due to gravity is given by;

$g = R/\Delta t^2$ (1)

The rate of change of energy P radiated as a result of the above acceleration will depend on the constants c (speed of light) and G (universal gravitational constant), hence;

$P = c^5/G$ (2)

The power and time must be re-quantized in units of $\hbar = h/2\pi$ where h is Planck constant, hence

$P\Delta t^2 = n^2\hbar$ (3)

Where n = 1,2,3........ is the principle quantum number.

But the total energy of the atom in the various energy states is $W = -ke^2/R$ where k is the Coulomb constant and e is the elementary charge. Since Δt^2 is known from Eqn1 and P from Eqn2 then using Eqn3 the radius is given by

$R = n^2 Gg\hbar /c^5$ (4)

From which the total energy is given by,

$W = -ke^2 c^5/ n^2 Gg\hbar$ (5)

From the Bohr-Einstein frequency (f) condition, applied to a transition from a level with n $=n_i$ to a level with n $= n_f$, The energy of a photon emitted by a hydrogen atom is given by the difference of two hydrogen energy levels

$hf = E_i - E_f$

Finally we get since frequency $f = c/\lambda$, where λ is the wavelength

$1/\lambda = [ke^2 c^4/ 2\pi G\hbar^2][1/g][1/ n_f^2 - 1/ n_i^2]$ (6)

The equation obtained above shows somehow a great significance of gravity in the quantum theory. So far it states that regardless of the levels in the transitions of an atom the acceleration due to gravity of the particles in the atom do greatly affect the nature of its spectrum.

The quantity $[ke^2 c^4/ 2\pi G\hbar^2]$ is the inverse of the square of time t and therefore; $1/t^2 = [ke^2 c^4/ 2\pi G\hbar^2]$, from which the time is obtained as t = 1.58873×10^{-42}s. This is the earliest period of time in the history of the universe.

Comparing Eq6 with Bohr's model, here we shall equate the Rydberg constant $[k^2 e^4 m/4\pi c\hbar^3]$, where m is the mass of the particle, to the constant $[ke^2 c^4/ 2\pi G\hbar^2][1/g]$. Doing this generates an acceleration given by $g_a = [$ $2\hbar c^5/ke^2 Gm]$, then at the Planck epoch when $m = \sqrt{\left(\frac{\hbar c}{G}\right)}$ the acceleration reduces to $g = \frac{8\pi \varepsilon_0}{e^2}\sqrt{\left(\frac{\hbar c^7}{G}\right)}$. Then At the Schwarz child's radius $R = Gm/c^2$ the acceleration is $g_b = c^4/Gm$ which gives an equation for the spectrum as $1/\lambda = [/ 2\pi a_0][1/ n_f^2 - 1/ n_i^2]$ where a_0 is the first Bohr radius $[\hbar^2/ mke^2] = 5.28 \times 10^{-11}$m.

The interesting part of it is that the ratio $g_b / g_a = [ke^2/2\hbar c]$ is the fine structure constant.

Chapter 13. Everything

Quantum gravity is the field of theoretical physics[1] that tries to unify quantum mechanics[2] with general relativity. Quantum mechanics describes the three fundamental forces of nature[3] while general relativity[4] is a theory of the fourth fundamental force: gravity[5]. The goal everyone is waiting for to emerge from this unification is a "theory of everything[6]", or "Grand Unified Theory[7]" (GUT). In 1986[8], Abhay Ashtekar[9] reformulated Einstein's field equations of general relativity using what have come to be known as Ashtekar variables[10], a particular flavor of Einstein-Cartan theory[11] with a complex connection. He was able to quantize gravity using gauge field theory[12]. In the Ashtekar formulation, the fundamental objects are a rule for parallel transport[13] and a coordinate frame known as a vierbein[14] at each point. Because the Ashtekar formulation was background-independent, it was possible to use Wilson loops[15] as the basis for a nonperturbative quantization of gravity. Explicit (spatial) diffeomorphism invariance of the vacuum state[16] plays an essential role in the regularization of the Wilson loop states.

---

1. http://en.wikipedia.org/wiki/Theoretical_physics

2. http://en.wikipedia.org/wiki/Quantum_mechanics

3. http://en.wikipedia.org/wiki/Fundamental_interaction

4. http://en.wikipedia.org/wiki/General_relativity

5. http://en.wikipedia.org/wiki/Gravitation

6. http://en.wikipedia.org/wiki/Theory_of_everything

7. http://en.wikipedia.org/wiki/Grand_Unified_Theory

8. http://en.wikipedia.org/wiki/1986

9. http://en.wikipedia.org/wiki/Abhay_Ashtekar

10. http://en.wikipedia.org/wiki/Ashtekar_variables

11. http://en.wikipedia.org/wiki/Einstein-Cartan_theory

12. http://en.wikipedia.org/wiki/Gauge_field_theory

13. http://en.wikipedia.org/wiki/Parallel_transport

14. http://en.wikipedia.org/wiki/Vierbein

15. http://en.wikipedia.org/wiki/Wilson_loop

16. http://en.wikipedia.org/wiki/Vacuum_state

Around 1990[17], Carlo Rovelli[18] and Lee Smolin[19] obtained an explicit basis of states of quantum geometry, which turned out to be labelled by Penrose's spin networks[20]. In this context, spin networks arose as a generalization of Wilson loops necessary to deal with mutually intersecting loops. Mathematically, spin networks are related to group representation theory and can be used to construct knot invariants[21] such as the Jones Polynomial.

The need for this chapter is to understand those problems involving the combination of very large mass or energy and very small dimensions of space, such as the behavior of black holes[22], and the origin of the universe[23]

The formula for the quantization of quantum gravity

The model is based on separating the gravitational field into the sum of two components; that is the background and the quantum field. The background left is one for all our calculations. But because loop gravity ignores the back ground space as a lost entity that does not occur in space, there fore the need to reconstruct quantum field theory from scratch without a background space is taken into account. I therefore suggest that the calculation should be performed by summing all possible space-times.

Quantum field theory[24] depends on particle fields embedded in the flat space-time of special relativity[25]. General relativity[26] models gravity as a curvature within space-time[27] that changes as a gravitational mass (m) moves. Assuming a spherical symmetric object that space time is of dimensions increasing from 1, 2, 3, 4...N, where N is the nth term of the dimensions. To quantize space and time is to create a space in which all of physics is quantized. The nature of the curved space surface is described by increasing powers in the

---

17.   http://en.wikipedia.org/wiki/1990

18.   http://en.wikipedia.org/wiki/Carlo_Rovelli

19.   http://en.wikipedia.org/wiki/Lee_Smolin

20.   http://en.wikipedia.org/wiki/Spin_network

21.   http://en.wikipedia.org/wiki/Knot_invariant

22.   http://en.wikipedia.org/wiki/Black_hole

23.   http://en.wikipedia.org/wiki/Big_Bang

24.   http://en.wikipedia.org/wiki/Quantum_field_theory

25.   http://en.wikipedia.org/wiki/Special_relativity

26.   http://en.wikipedia.org/wiki/General_relativity

27.   http://en.wikipedia.org/wiki/Spacetime

Schwarzschild radius $R_s = Gm/c^2$, Hence describing the dimensions of space. Quantum mechanics explains the existence of discrete energy states in an atom, in away that the angular momentum of the atom must be quantized, which is also the case for quantum gravity. The equation for the quantization of the loop quantum gravity can then be written as,

$$\eta R_s + \beta R_s^2 + \mu R_s^4 + \ldots\ldots\ldots + \delta R_s^N = n\hbar \;[1]$$

Where $\eta = \sqrt{Beh}$, is the momentum of a particle probing another form of quantum mechanics, $\hbar = h/2\pi$, where h is Planck constant, $\beta = 8\pi Be$, e is the elementary charge, B is the magnetic field and finally $\mu = 256\pi^3 P/c^2$, where P is the intensity and c is the constant speed of light.

The energy equation

What changes is the form of the equation the rest remaining constant. The principle behind this is that eqn1 can be changed to any form simply for purposes of calculating complex phenomenon. The energy to which we are concerned here is expressed as a general expression describing the energy scales forming smaller and larger matter entities in the universe. The energy will thus be given by;

$$\eta c + \beta c R_s + \mu c R_s^3 + \ldots\ldots\ldots + \delta c R_s^{N-1} = n\hbar c/R_s \;[2]$$

Note: the background space described by the Schwarzschild radius has changed, thus the above equation in any case can be used to calculate the basic properties of Black holes. Remember the Schwarzschild radius is the radius for a given mass where, if that mass could be compressed to fit within that radius, no known force or degeneracy pressure could stop it from continuing to collapse into a gravitational singularity[28].

The mass equation

Having explored the energy scale we now form general equation that describes well the mass scale. This is also done the same way as eqn2 and therefore generate,

$$\eta/c + \beta R_s/c + \mu R_s^3/c + \ldots\ldots\ldots + \delta R_s^{N-1}/c = n\hbar/c R_s \;[3]$$

The maximal magnetic field

Assuming that the energy $W = \beta c R_s$, from eqn2 is equal to the energy $W = mc^2$, we hence obtain the magnetic field as, $B = c^3/8\pi Ge = 1.0054 \times 10^{53}$ N/

---

28.   http://en.wikipedia.org/wiki/Gravitational_singularity

Am. using this magnetic field in the energy equation, $W = \eta c$ we get the energy in the form $W = (c^2/2) \sqrt{\hbar c/G}$ where the quantity $\sqrt{\hbar c/G}$ is the Planck mass M_p at an energy of 6.119×10^{18} GeV.

Time taken by a black hole to evaporate and its entropy

The energy required here is given in Eqn2, it is at this, that the intensity $P = W/A\Delta t$, (where A is the area and t is the time) is used. We take the energy $W = \mu c R_s^3$ (from Eq2) as our interest from which we obtain the time as $\Delta t = 256\pi^3 R_s^3/Ac$. But with black holes the area will become exactly equal to the square of the Planck length as $A \sim L_p^2 = \hbar G/8\pi c^3$ hence the change in time is given by $\Delta t = 63500.86\pi\, G^3 m^3/\hbar c^4$.

For entropy we set the energy to kT, where k is Stefan's-Boltzmann's constant and T is the temperature of the body. Now for $kT = \mu c R_s^3$, since Δt is known the entropy is thus given by $S = W/T = 78.96 A k c^3/\pi\hbar G \sim A/4$. In conclusion we state that the entropy of a black hole is proportional to the area of the event horizon.

The quantum Hall Effect

For this effect the momentum η is used. From Eqn2 we set, $\eta c = n\hbar/R_s$ which gives the magnetic flux as $4\pi R_s^2 B = nh/e$, from which the resistance is given by $\zeta = 4\pi R_s^2 B/e = nh/e^2$. for n= 1,2,3,4 the resistance is of a value 25833.8Ω.

Maximum Intensity

Using eqn3 in this case, since B is known and P got from $\mu R_s^4 = n\hbar$; as $P = \hbar c^2/256\pi^3 R_s^4$, we hence obtain, $M_p/2 + m + M_p/m = M_p/m$, which gives $M_p + 2m = 0$, and for identical mass M = 0, which is true. The intensity at the planck length that is for $R_s = L_p$ is $P = c^8/\pi\hbar G^2$

Chapter 14. The Theory of Light

Basing our study on the electric currents generated whenever there is a changing magnetic field (B) and a changing electric field (E) in the electromagnetic wave we can construct a complete theory for the electromagnetic radiations. The theory is created using the symmetry between a long wire placed in the electromagnetic fields which induce vibrating electrons that carry current in the wire and the electromagnetic wave which constitute changing electric and magnetic fields that create vibrating photons in the wave. Therefore a wire is to a wave what a vibrating electron is to a vibrating photon in the wire and a wave respectively. The aim of the paper is to give a clear description of the theory of electromagnetic radiations (light). The goal of the paper on the other hand is to show that the wave-particle descriptions of reality can be applied to any physical situation simultaneously. The objective of the paper is to show that the Photoelectric Effect and the Compton Effect can both be explained by the wave model and the particle model at the same time.

Consider a long wire connected to an ammeter and strong electric and magnetic fields produced in a vacuum. Let us assume that whenever a wire is brought in vicinity of a changing electric field, electrons of mass (m) are set into motion in the wire and then an ammeter deflects, recording a current (i_E). The current in the wire due to a changing electric field should be given by

$$i_E = \frac{j\varepsilon_0}{2\pi m} E \ (1)$$

Where (ε_0) is the permittivity of free space and (j) is the constant of action in SI units Js. therefore the current is quantized and depends on both the electric field and the mass of an electron.

When the wire is brought into the magnetic field, vibrating electrons at a frequency of oscillation (f) are set in motion at a speed (v) through the wire generating a current given by

$$i_B = \frac{v}{2\pi \mu_0 f} B \ (2)$$

Where (μ_0) is the permeability of free space.

Assuming that the ammeter records different values of ($^i E$) and ($^i B$), what will be the change in the current values recorded at the ammeter? Subtracting equation (1) from equation (2) we have

$$\Delta I = (i_E - i_B) = \left(\frac{j\varepsilon_0}{2\pi m} E - \frac{v}{2\pi \mu_0 f} B \right) \tag{3}$$

This is the change in the currents due to changing magnetic and electric fields. Assuming that there is no change in the current, meaning that the current values for $^i E$ are equal to those of $^i B$ (i.e $\Delta I = 0$). This will imply that the magnetic field strength was equal to the electric field strength at one point in both experiments. In terms of electromagnetic radiations in the vacuum, assuming that a wire carrying current is replaced by a wave and electrons are replaced by photons. The wire replaced by a wave is made up of vibrating electric and magnetic fields at a given frequency making an electromagnetic wave. The electrons replaced with photons will represent the particle properties of the electromagnetic wave (light) with associated mass and speed (v).

The symmetry here is between the long wire and the wave, the electrons and the Photons. The electric and magnetic fields brought in vicinity of the wire and the number of oscillations per second of the electron in the wire is what leads to an electromagnetic wave. The electrons with a given mass and moving at a given speed is what constitute a photon. Then at $\Delta I = 0$, we have on arranging,

$$\frac{jf}{mv} = \frac{1}{2\pi \mu_0 \varepsilon_0} \frac{B}{E} \tag{4}$$

This means that at $\Delta I = 0$, either a changing magnetic field or a changing electric field produces a current. Then it should be true that a changing magnetic field produces an electric field just as a changing electric field produces a magnetic field. This process in the electromagnetic wave continues indefinitely. The electromagnetic wave will move at a constant speed (c), since for electromagnetic waves, $\frac{E}{B} = c$, and for a photon $\frac{jf}{mv} = c$ where j=6.63× 10^{-34} Js (also called the Planck constant after Max Planck) and mv is the photon momentum. Implying that the photon energy is related to the frequency of the electromagnetic wave by (jf). Then the electromagnetic wave will move at a constant speed given as, since by symmetry $\frac{E}{B} = \frac{jf}{mv} = c$

$$c = \frac{1}{\sqrt{\varepsilon_o \mu_o}} = 2.99792458 \times 10^8 \, \frac{m}{s}$$

Where $\varepsilon_o = 8.85418782 \times 10^{-12} \, \frac{c^2}{Nm^2}$ and $\mu_o = 1.26 \times 10^{-6} \frac{Ns^2}{c^2}$

We have therefore deduced based on the symmetry between a current (electron) carrying wire in the electromagnetic field and the photons in electromagnetic waves that an electromagnetic wave moves at a constant speed of light. It is also true from the deductions that light is indeed made up of particles of light called photons and vibrating electric and magnetic fields. The deduction would not be possible if the wave and particle descriptions of the situations had not been applied simultaneously (into what is called "the wave-particle duality).

Unexpectedly enough the photoelectric effect can also be explained by Equation (3), on arranging

$$\frac{2\pi m f}{\varepsilon_o E} \Delta I = jf - \frac{mv}{2\pi \mu_o \varepsilon_o} \frac{B}{E}$$

Then the total energy of the particle of light (Photon) is then given by

$$jf = \frac{2\pi m f}{\varepsilon_o E} \Delta I + \frac{mv}{2\pi \mu_o \varepsilon_o} \frac{B}{E} (5)$$

It is therefore true that the photoelectric effect can be explained when both the particle and wave models of reality are applied in the experiment at the same time (simultaneously). The work function from Einstein's photoelectric equation (A. Einstein, 1905) will here be replaced by $\frac{2\pi m f}{\varepsilon_o E} \Delta I$ while the kinetic energy of the electrons at the surface of the metal will be given by $\frac{mv}{2\pi \mu_o \varepsilon_o} \frac{B}{E}$. Equation (5) reduces to Einstein's Photoelectric effect when, the speed of the electron is $v = \frac{1}{\pi \mu_o \varepsilon_o} \frac{B}{E}$ and the change in current for a complete circuit is $\Delta I = \frac{j \varepsilon_o E}{2\pi m}$.

The validity of the Compton Effect can also be deduced from Equation (3). The current can be taken as the product of the frequency (f) of radiations and the charge (q) on the particle.

Then the current due to the electric field is $i_E = qf_1$ and that due to the magnetic field is $i_B = qf_2$. In the case of the Compton Effect, q is the charge on the free electron while f_1 and f_2 are the frequencies of the incoming photon and outgoing photon after collision with the free electron respectively. Then equation (3) can be written as

$$f_1 - f_2 = \frac{1}{q}\left(\frac{j\varepsilon_0}{2\pi m}E - \frac{v}{2\pi \mu_0 f}B\right) \quad (6)$$

Since photons move with the speed of light(c) then their frequencies is related to their speed and wavelength by $f = \frac{c}{\lambda}$, then we have

$$\frac{1}{\lambda_1} - \frac{1}{\lambda_2} = \frac{1}{qc}\left(\frac{j\varepsilon_0}{2\pi m}E - \frac{v}{2\pi \mu_0 f}B\right)$$

On arranging to include the charge density of the free electron for electric field lines in an area of $\frac{\lambda_1 \lambda_2}{2\pi}$, we obtain

$$\frac{2\pi q}{\lambda_1 \lambda_2}(\lambda_2 - \lambda_1) = \frac{j}{mc}\left(\varepsilon_0 E - \frac{mv}{\mu_0 jf}B\right)$$

Where (mc) is the momentum of an electron treated relativistic ally, on letting the charge density $= \frac{2\pi q}{\lambda_1 \lambda_2} = \varepsilon_0 E$, we deduce the change in the wave length of the incoming photon and outgoing photon after collision with the free electron as

$$\Delta\lambda = (\lambda_2 - \lambda_1) = \frac{j}{mc}\left(1 - \frac{mv}{\rho\mu_0 jf}B\right)$$

Since $\rho = \varepsilon_0 E$, we then have

$$\Delta\lambda = (\lambda_2 - \lambda_1) = \frac{j}{mc}\left(1 - \frac{\frac{mvB}{\varepsilon_0 \mu_0 E}}{jf}\right)$$

Since jf is the energy carried by the photon, and then also $\frac{mvB}{\varepsilon_0 \mu_0 E}$ is the energy carried by the free electron. Treating the electron relativistically such that for electromagnetic waves moving at a speed (v) relative to the electron moving at a speed of light $c = \frac{1}{\sqrt{\varepsilon_0 \mu_0}}$, the electric field in the wave will be related to the magnetic field by $Bv = E.$ then the energy carried by an electron can be given by mc^2. Then the angle at which the photon is scattered after collision with the free electron will be given by

$$\theta = \cos^{-1}\left(\frac{mvB}{\varepsilon_0 \mu_0 E}\right)\Big/ jf$$

$$(7)$$

Where mv is the momentum of the photon in the electromagnetic wave consisting of a changing electric field E and magnetic field B both moving at a constant speed of light $c = \frac{1}{\sqrt{\varepsilon_0 \mu_0}}$. Treating the electron relativistically we have $\theta = \cos^{-1}\frac{mc^2}{jf}$

When the energy carried by the photon is equal to the energy possessed by the electron then $\theta = 0$, meaning that there is or there is no scattering and whatsoever there is no increase in photon wavelength hence $\Delta\lambda = 0$.

A complete theory of light can't fail to explain the structure of an atom. I therefore take a complete discussion of what goes on inside an atom only with the help of Bohr's energy levels which he derived using classical mechanics and quantum theory. Let $\Delta f = f_1 - f_2$ be an increase in the frequency of the electromagnetic radiations emitted from an atom. Then squaring both sides of equation (6) and arranging will give

$$4\pi^2\Delta f^2 = \frac{1}{m^2 q^2}\left(j\varepsilon_0 E - \frac{mv}{\mu_0 f}B\right)^2$$

$$4\pi^2 m^2 q^2 \Delta f^2 = j^2\varepsilon_0^2 E^2 - 2\frac{j\varepsilon_0 EBmv}{\mu_0 f} + \frac{B^2 m^2 v^2}{\mu_0^2 f^2}$$

Dividing through by $64\pi^4 j^2\varepsilon_0^2$ and multiplying through by q^2 gives the energy of the atom as on arranging

$$\frac{mq^4}{16\pi^2 \pi^4 j^2 \varepsilon_0^2} = \frac{1}{64\pi^4 m\Delta f^2}\left((Eq)^2 - 2\frac{m(Eq)(Bqv)}{\mu_0\varepsilon_0(jf)} + \frac{(Bqv)^2 m^2}{\mu_0^2\varepsilon_0^2(jf)^2}\right)$$

The energy of the n-th level is since the reduced Planck constant is $n\hbar = \frac{nj}{2\pi}$

$$\frac{mq^4}{32\pi^2 \pi^4 n^2\hbar^2 \varepsilon_0^2} = \frac{1}{32\pi^2 n^2 m\Delta f^2}\left((Eq)^2 - 2\frac{m(Eq)(Bqv)}{\mu_0\varepsilon_0(jf)} + \frac{(Bqv)^2 m^2}{\mu_0^2\varepsilon_0^2(jf)^2}\right)$$

The expression on the left hand side of the equation is the quantized energy of an atom (Niels Bohr, 1913) while the right hand side of the equation represents the energy of the atom in terms of the forces associated with it. In the equation we let $H_e = Eq$ be the electric force for a particle moving in the electric field and $H_b = Bqv$, the magnetic force on a particle with charge q moving in the magnetic field. Since the speed of light is $= \frac{1}{\sqrt{\varepsilon_0 \mu_0}}$, then the quantized energy can be given as

$$W_n = \frac{1}{32\pi^2 n^2 m \Delta f^2} \left(H_e{}^2 - 2\frac{H_e H_b mc^2}{jf} + \frac{H_b{}^2(mc^2)^2}{(jf)^2} \right)$$

Then on arranging we obtain

$$W_n = \frac{1}{32\pi^2 n^2 m \Delta f^2} \left(H_e - \frac{mc^2}{jf} H_b \right)^2 \quad (8)$$

When the energy of an electron moving at a speed of light in atom is equal to the energy of the emitted photon, then

$$W_n = \frac{1}{32\pi^2 n^2 m \Delta f^2} (H_e - H_b)^2 = \frac{1}{32\pi^2 mn^2} \left(\frac{\Delta H}{\Delta f} \right)^2 \quad (9)$$

Where $\Delta H = H_e - H_b$ is the difference or change between the electric force and the magnetic force in an atom, when the two forces balance (i.e. $H_e = H_b$), then $W_n = 0$ meaning that the total energy of an atom will cease to exist.

Therefore the total energy of an atom increases with the square of the change in the electric and magnetic forces which govern an electron but falls off as the square of the change in the frequency of the radiation emitted by it.

From equation (8) the ratio of the energy of an electron to that of the photon $\frac{mc^2}{jf}$, is the limit at which if the energies are not equal you will not get a change in the electric and magnetic forces. Treating the ratio as a number $\tau = \frac{mc^2}{jf}$, we get from equation (8)

$$W_n = \frac{1}{32\pi^2 mn^2} \left(\frac{H_e - \tau H_b}{f_1 - f_2} \right)^2 \quad (10)$$

When $\tau = 0$, it means that the relativistic energy (mc^2) of an electron in an atom is zero, and that the total energy of an atom only increases with the electric force on the electron. The relationship (equation 10) is a complete expression for the laws according to which, by the theory here advanced, the structure of an atom should be viewed.

In conclusion, a complete theory of light is only possible if both the wave and particle descriptions of reality are applied to the physical situation at the same time. In discussing Young's double slit experiment for example we should be able with the formulas given above to treat the electromagnetic radiations on both a wave and particle model.

Chapter 15. The Bekenstein-Hawking Area-Entropy Law

The Unification of Quantum Mechanics and General Relativity into a Quantum theory of Gravity is one of the great scientific challenges of this generation. A definitive resolution will require solving one of the major problems of Quantum Gravity and that is, the Bekenstein-Hawking area-entropy law, $S = a\frac{Ac^5 k}{\hbar G}$ (1), where A is the surface area of the Schwarzschild black hole, a is the constant of the order of unity, c is a constant speed of light, k the Boltzmann constant, $\hbar$ the reduced Planck constant and G is the Newton's gravitational constant.

Attempts towards this were done in the early 70s by Hawking who proved that a black hole emits thermal radiation with a temperature $T = \frac{\hbar c^3}{8\pi G k}$ (2). According to Carlo Rovelli (Dec, 2003), Hawking beautiful result raises a number of questions. First, in Hawking's derivation the quantum properties of gravity are neglected. Are these going to affect the result? Second, we understand macroscopical entropy in statistical mechanical terms as an effect of the microscopical degrees of freedom. What are the microscopical degrees of freedom responsible for the entropy? Can we derive (1) from first principles? Because of the appearence of $\hbar$ in (1), it is clear that the answer to these questions has since become standard benchmark against which a quantum theory of gravity can be tested.

This book presents a simple universal explanation of Black hole thermodynamics in a somewhat different form than that given by Loop Quantum Gravity (LQG), String theory and Hawking radiation theory. The major result of the book is the derivation of (1) from first principles using different methods for Schwarzschild and for other black holes, with a well defined calculation where no infinities appear. As far as this book is concerned there is no other theory from which such a calculation can proceed. Hence the book is the only one from which a detailed quantum theory of gravity precedes and where the result of the Bekenstein-Hawking area entropy law can be achieved.

In this method we reduce the famous Einstein field equation ($G_{\mu\nu} + \Lambda g_{\mu\nu} = \frac{8\pi G}{c^4} T_{\mu\nu}$ where, the expression on the left represents the curvature of space time while the expression on the right represents the matter-energy content of the universe) to,

$$\frac{1}{R^2} = \frac{8\pi G}{c^4} P_{eg} \quad (3)$$

Where, R is the radius of a body of mass M, $P_{eg} = \sigma_m \frac{f_e f_g}{\hbar c}$ is the Pressure-Energy density relationship with the coupling of mass (the ratio of the atomic mass, m to the Planck mass M_{pl}) and the electric f_e and gravitational force f_g .

The ratio, $\sigma_m = \frac{m}{M_{pl}}$ is introduced to correct for particles approaching the Planck length scale $m \to M_{pl}$

What is the total electric potential energy of a black hole? From (3), we could let the potential electric energy be,

$$E_e = f_e r = \frac{\hbar c^5}{8\pi G E_g \sigma_m}$$

We know that at the Schwarzichild radius $R = \frac{GM}{c^2}$, the gravitational potential energy will be of order $E_g = mc^2$, giving the electric energy from (1) as,

$$E_e = \frac{\hbar c^3}{8\pi G M \sigma_m}$$

What is the temperature of a Black hole? Since the thermal energy is given by $E_{thermal} = kT$, where k is the Boltzmann constant

By the principal of Equipartition

$$E_{thermal} \sim E_e \Rightarrow T = \frac{\hbar c^3}{8\pi G M k \sigma_m} \quad (4)$$

For $\sigma_m = 1$, we get the usual Hawking temperature of

$$T = \frac{\hbar c^3}{8\pi G M k} \quad (5)$$

We know that, entropy is energy divided by temperature. Having derived the temperature, What is the total energy of a Black hole?

Assuming a law which states that the intensity of the emitted radiation increases as the square of the electric force $I = \beta F_e^2$, where the constant $\beta = \frac{1}{4h}$,

But we can also write the intensity in terms of energy as, $I = \frac{E_T}{tA}$, where t is time and A is the surface area of Schwarzschild black hole. The total energy of a Black hole will then be given as,

$$E_T = \frac{F_e^2 tA}{4h}$$

Let the time taken by a Black hole to evaporate be, $= \frac{Mc}{F_e}$, F_e is known from (3), since from the Newtonian law of gravity $F_g R^2 = GM^2$ we then have the total energy of a black hole as,

$$E_T = \frac{Ac^6}{32\pi\sigma_m G^2 M} (6)$$

And Power is given by $P = F_e c = \frac{hc^6}{8\pi\sigma_m G^2 M^2}$

Then the entropy of a Black hole is given by

$$S = \frac{E_T}{T}$$

Substituting in (6) and (4) we obtain the Bekenstein-Hawking area entropy law,

$$S = \frac{Ac^3 k}{4hG} (7)$$

Where the constant a=1/4

Chapter16. Making Sense with Semi-Classical Gravity

For the past thirteen years, I have been deriving the most important theories of physics from scratch without employing the methods of general relativity and quantum field theories and I have come up with promising results. I have deduced the Black hole thermodynamics from first principles, I have deduced the Wiedmann Franz law from scratch, the Stefan Boltzmann power law, The result for the earliest period of time in the history of the Universe, I have related the Chandrasker theory of white dwarfs with the Bohr theory of the Hydrogen atom-the results are suprising, the rest is history. This book gives a clear account of these fields of physics.

The truth is, I hate Einstein and Hawking. I don't like them because I find it hard to use their mathematical ideas to deduce the theories I desire. It was that hard for me to classify where in the scientific community I fall, at first I thought that my ideas where into the quantum gravity field section but this was a lie. The quantum theory of gravity has not been fully settled. It was yesterday that I realized that my ideas fell into the Semi-Classical physical regime when I browsed it online;

"Semi-classical physics refers to a theory in which one part of a system is described quantum-mechanically whereas the other is treated classically" In general, it incorporates a development in powers of Planck's constant, resulting in the classical physics of powers 0, and the first nontrivial approximation into the powers of -1. (Wikipedia)

I am sorry, Semi-classical physics hasn't gained much interest, there are too many criticism about its meaning, researches into the field have been discouraged, few physicists have written about it and it is that unimportant. But anyway I am an amateur to venture into a field that is irrelevant. I don't give a damn what you think.

My first insight into the field of Semi- classical physics is traced back in 2010 in my first paper I published on arXiv.org titled "A hypothetical investigation into the realm of the microscopic and macroscopic universes beyond the standard model" This paper clearly shows that I was into the field without knowing. For sure I thought I was dealing with the field of Quantum Gravity by then.

Well, if you don't understand Semi-classical physics, Amateurs do. Below I show you why I think I understand the field and you surely do. I provide many ideas which I think the entire scientific community must investigate.

The meaning of semi-classical physics to an amateur

Assuming an experiment where the classical electric force f_e is balanced over the classical gravitational force f_g to determine their strength, the result will show that, the ratio of the two forces will follow a power law in powers of n of the gravitational coupling constant as,

$$\frac{f_e}{f_g} = \alpha_g{}^n$$

The left hand side of the equation represents the classical part of the system while the right hand side represents the quantum mechanical part of the system.

Let the classical part be described by two constants;

G-The Universal gravitational constant

c- The constant speed of light

Into (G, c)

Let the Quantum mechanical part be described by two constants,

$\hbar$- The reduced Planck constant

c-The constant speed of light

Into ($\hbar$, c)

Then from the above assumption Semi-classical physics will reduce results combining the constants above into (G, c, $\hbar$)

From the above formula we can deduce the time and length units of measure formulas to help us understand the field better,

Time $t_n = \frac{Gm}{c^3} \alpha_g{}^{-n}$

Length $l_n = \frac{Gm}{c^2} \alpha_g{}^{-n}$

Where m denotes the mass of a particle or body and $\alpha_g = \frac{Gm^2}{\hbar c}$ is the gravitational coupling constant. You can also assume interactions involving the electromagnetic coupling constant.

The different fields of physics resulting from the above classification for different powers of n from 0, 1, 2 and
-1/2 are given below,

For n=0

<u>Classical General Relativity</u>

$$t_0 = \frac{Gm}{c^3}$$

$$l_0 = \frac{Gm}{c^2}$$

For n=1

<u>Quantum mechanics</u>

$$t_1 = \frac{\hbar}{mc^2}$$

$$l_1 = \frac{\hbar}{mc}$$

For n=2

<u>Semi-classical gravity</u>

$$t_2 = \frac{\hbar^2}{Gcm^3}$$

$$l_2 = \frac{\hbar^2}{Gm^3}$$

For n= -1/2

<u>Planck Units</u>

$$t_{-1/2} = \left(\frac{G\hbar}{c^5}\right)^{1/2}$$

$$l_{-1/2} = \left(\frac{G\hbar}{c^3}\right)^{1/2}$$

The above derivation gives out a clear description of Semi-classical physics to a lay person.

One can decide to use n as a spatial dimension of space.

Applications of semi-classical physics

Radiation intensity of a black hole

<u>The classical part of a system</u>

Let the classical total force on an electron in orbit at a distance r from the nucleus of an atom be related to its electromagnetic and gravitational forces by,

$$f = \frac{F_G F_e}{F_B}$$

Where F_G is the gravitational force, F_e is the electric force and $F_B = Bev$ is the magnetic force

The angular momentum of an electron is given classically as,

$$L = \frac{Gm^2}{c} = mvr$$

The Quantum mechanical part of the system

The angular momentum is quantized as,

$$L = \frac{K_e e^2}{c} = \hbar$$

On eliminating the constant speed of light c from both the expression of the angular momentums we have

$$mvr = \frac{F_G}{F_e} \hbar$$

The ratio $\frac{F_G}{F_e}$ represents the classical part of the system while $\hbar$ represents the quantum part.

Eliminating F_G from the above expression we get the magnetic power as,

$$F_B c = \frac{2\pi r^2 \lambda mv F_e^2}{h^2}$$

But the de Brogile wave length of an electron is,

$\lambda = h/mv$

And the surface area of the sphere of orbit of an electron is A $=4\pi r^2$.

Then the electromagnetic Intensity is given as,

$$I = \frac{F_B c}{A} = \frac{F_e^2}{2h}$$

Thus the intensity of a wave is proportional to the square of the electric force If we let the power of the electromagnetic wave be P= F$_B$c, and n be the fine structure constant $\alpha = ke^2/\hbar c$, then the equation for the intensity of the classical electromagnetic wave comes out clearly as, P = EB /μo$=2\varepsilon_0 E^2 c$, Where μo is the permeability of free space.

The intensity of radiations of black holes

From our previous expression

$$\frac{f_e}{f_g} = \alpha_g{}^n$$

At n = -1, and $f_g = \frac{c^4}{8\pi G}$ we have the electric force as,

$$f_e = \frac{\hbar c^5}{8\pi G^2 m^2}$$

Then the intensity of the radiations will be given as

$$I = \frac{f_e{}^2}{2h} = \frac{\hbar c^{10}}{256\,\pi^2 G^4 m^4}$$

This expression comes from treating the particle classically in one part and then quantum mechanically in another part.

It can be clearly seen above, that we haven't used the mathematics of general relativity or quantum field theory to reach at the result.

The earliest period of time in the history of the universe

Classical part of the system

Let the acceleration due to gravity of a particle (say an electron) in the gravitational field be given as

$g = R/\Delta t^2$

Where is Δt the time

And R is the distance of the particle from the source

If the particle radiates energy then the energy per unit time is,

$P = c^5/G$

Quantum mechanical part of the system

The power and time must be quantized in units of $\hbar = h/2\pi$ where h is Planck constant, hence

$P\Delta t^2 = n^2 \hbar$

Where n= 1,2,3........ is the principle quantum number.

But the potential energy of the electron in the various energy states is W= $-ke^2/R$ where k is the Coulomb constant and e is the elementary charge. Since Δt^2 is known from the expression for acceleration due to gravity.

Then the distance R is,

$R = n^2 Gg\hbar /c^5$

From which the total energy is given by,

$W = -ke^2 c^5/ n^2 Gg\hbar$

From the Bohr-Einstein frequency (f) condition, applied to a transition from a level with n =n_i to a level with n = n_f, The energy of a photon emitted by a hydrogen atom is given by the difference of two hydrogen energy levels

$hf = E_i - E_f$

Since frequency $f = c/\lambda$, where λ is the wavelength

Then we have

$$1/\lambda = [ke^2c^4/2\pi G\hbar^2][1/g][1/n_f^2 - 1/n_i^2]$$

The equation obtained above shows some how a great significance of gravity in the quantum theory. So far it states that regardless of the levels in the transitions of an atom the acceleration due to gravity of the particles in the atom do greatly affect the nature of its spectrum.

The quantity $[ke^2c^4/2\pi G\hbar^2]$ in the formula above is the inverse of the square of time t and therefore ,

$$1/t^2 = [ke^2c^4/2\pi G\hbar^2],$$

From which the time is obtained as t = 1.58873×10^{-42}s. This is the earliest period of time in the history of the universe.

The Weidmann Franz- Lorenz law

Treating one part of the system classically (macroscopic) and the other quantum mechanically (microscopic), we have the formula for the electric force acting on an electron in motion as

$$F = \frac{n^2}{a_g}f_g$$

Where n, is the principle quantum number.

The above formula differs from the one previously given

On squaring the above equation we obtain the square of the electric field as,

$$E^2 = \frac{n^4c^4}{G^2e^2m^2}\left(\frac{c^3\hbar}{8\pi Gm}\right)^2$$

From the formula for the temperature of the black hole, the function $\frac{c^3\hbar}{8\pi Gm}$ is related to temperature as kT, and then the law for thermal conductivity will be reduced as,

$$\frac{\pi^2E^2G^2m^2}{3Tc^4} = \left(\frac{n^4\pi^2}{3}\right)\left(\frac{k}{e}\right)^2 T$$

The left hand side represents the ratio of the thermal conductivity K to the electric conductivity δ. The right hand side is the Weidman –Franz law. Therefore the left side of the equation represents the macroscopic part of the system while the right hand side represents the microscopic part of the system.

Then the left-hand side will be given as,

$$\frac{K}{\delta} = \frac{1}{3}\left(\frac{\pi Gm}{c^2}\right)^2 \frac{E^2}{T} = \frac{\pi A}{3}\frac{E^2}{T}$$

Where A is the surface area of a body $A = \pi r_s^{\ 2}$ with the schwarzichild's radius r_s. This is the conductivity ratio of a black hole.

Chapter 17. The Art of Reductionism

Scientific reductionism is the idea of reducing complex interactions and entities to the sum of their constituent parts, in order to make them easier to study (explorer.com). It is based on the idea that science can be used to explain everything by a mere look at the individual constituent processes.

There are three types of reductionism, that is, ontological, methodological and theory reduction. In this section we shall emphasize theory reduction because we have a great deal of reducing known laws of physics from a somewhat simple rule. This was the case when Kepler's laws of the motion of planets and Galileo's theories of motion were reduced to the Newtonian theories of mechanics.

Newtonian Mechanics became a more general theory simply because all the explanatory power of Kepler's and Galileo's laws was contained in it. Therefore theoretical reduction is considered as the reduction of one explanation or theory to another.

The most interesting thing about this section is that, during the process of reduction we create a relationship between the known law to another law explaining the same thing but unknown to the entire physics community. For example in the reduction of the Weidman Franz- Lorenz law we create in a process a law for the thermal conductivity of gravito-electric phenomenon.

Therefore reductionism is deriving something complicated from something simple. For example in the derivation of the Weidman Franz law we set a formula that states that, the electric force (Ee) on an electron is proportional to the gravitational force at the schwarzichild's radius $\left(\frac{c^4}{8\pi G}\right)$ but inversely proportional to the gravitational coupling constant $\left(\frac{Gm^2}{\hbar c}\right)$ as given below,

$$F = \frac{n^2}{a_g} f_g$$

Where n, is the principle quantum number.

On squaring the above equation we obtain the square of the electric field as,

$$E^2 = \frac{n^4 c^4}{G^2 e^2 m^2} \left(\frac{c^3 h}{8\pi G m}\right)^2$$

From the formular for the temperature of the black hole, the function $\frac{c^3 \hbar}{8\pi G m}$ is related to temperature as kT, and then the law for thermal conductivity will be reduced as,

$$\frac{\pi^2 E^2 G^2 m^2}{3Tc^4} = \left(\frac{n^4 \pi^2}{3}\right)\left(\frac{k}{e}\right)^2 T$$

The left hand side represents the ratio of the thermal conductivity K to the electric conductivity δ, which is the Weidman Franz law. From the above reduction we have generated an important rule given by,

$$\frac{K}{\delta} = \frac{1}{3}\left(\frac{\pi G m}{c^2}\right)^2 \frac{E^2}{T} = \frac{\pi^2 r_s^2}{3} \frac{E^2}{T}$$

The above formula explains the thermal properties of black holes at the schwarzichild's radius r_s.

Chapter 18. Construction of a Consistent Physical Theory of Nature

A consistent theory of nature, simply the "theory of everything" is constructed using one of the profound ideas of "axioms", that, when the Stoney units of measure are multiplied by the coupling constant (a dimensionless number) of a form $\alpha^{\frac{n-1}{2}}$, one can easily calculate the mass of all particles in the universe and their length of time scales with accuracy provided the value of n is known. The mass of the electron is calculated at $\alpha=1/137.036$ and n=21.32 while the mass of the earth is calculated at n= -29.99, hence solving one of the major unsolved problems in physics. The Planck mass is calculated and determined in principle to be 5.4556×10^{-8} Kg, a different value from the given value would lead to variations in our fundamental physical constants of electricity and gravity. The energy scales at given length scales in literature are also deduced in which a requirement to revisit our profound known physical theories is proposed.

One of the major unsolved problems in physics is developing a final theory, ultimate theory or theory of everything. In this paper we present a series of hypotheses and speculations leading inescapably to a conclusion that when the Stoney fundamental units of measure are multiplied by the electromagnetic coupling constant (fine structure constant) powered by any integer, $\alpha^{\frac{n-1}{2}}$ from 0,1,2,.....................,n, one gets to calculate the mass of all particles in the universe, the lengths between them and the time expressible at a scale of the known fundamental physical constants of nature. Our hypotheses may be wrong and our speculations idle, but the uniqueness and simplicity of our scheme are reasons enough that it be taken seriously.

Our starting point is the assumption that all of the fundamental physical units of measure can be calculated and organized to demonstrate different branches and scales of physics whatsoever using the following formulas,

Length

$$L_n = \frac{e}{c^2} \sqrt{\frac{G}{2\varepsilon_0}} \, \alpha^{n-1} \quad (1)$$

Time

$$t_n = \frac{e}{c^3}\sqrt{\frac{G}{2\varepsilon_0}}\,\alpha^{n-1} \quad (2)$$

Mass

$$M_n = e\sqrt{\frac{1}{2G\varepsilon_0}}\,\alpha^{n-1} \quad (3)$$

Where α is the coupling constant for either electromagnetic or gravitational interactions, G is the universal gravitational constant, e is the elementary charge on an electron, c is the speed of light and ε_0 is the permittivity of free space, the meaning of n is left to be investigated as per the meaning of the theory.

case1:

We derive the fundamental units of measure at values of n=0, 1,2,3,4 and 5 only for the fine structure constant $\alpha = \frac{e^2}{4\pi\varepsilon_0\hbar c}$ where is the reduced Planck constant $\hbar = \frac{h}{2\pi}$.

At n=0

$$L_o = \sqrt{\frac{2\pi G\hbar}{c^3}}, \quad t_o = \sqrt{\frac{2\pi G\hbar}{c^5}}, \quad M_o = \sqrt{\frac{2\pi\hbar c}{G}}$$

At n=1

$$L_1 = \frac{e}{c^2}\sqrt{\frac{G}{2\varepsilon_0}}, \quad t_1 = \frac{e}{c^3}\sqrt{\frac{G}{2\varepsilon_0}}, \quad M_1 = \frac{e}{\sqrt{2G\varepsilon_0}}$$

At n=2

$$L_2 = \frac{e^2}{\varepsilon_0}\sqrt{\frac{G}{8\pi c^5 \hbar}}, \quad t_2 = \frac{e^2}{\varepsilon_0}\sqrt{\frac{G}{8\pi c^7 \hbar}}, \quad M_2 = \frac{e^2}{\varepsilon_0}\sqrt{\frac{1}{8\pi G\hbar c}}$$

At n=3

$$L_3 = \frac{e^3}{\pi c^3}\sqrt{\frac{G}{32\varepsilon_0^3}}, \quad t_3 = \frac{e^3}{\pi c^4}\sqrt{\frac{G}{32\varepsilon_0^3}}, \quad M_3 = \frac{e^3}{\pi\hbar c}\sqrt{\frac{1}{32\varepsilon_0^3 G}}$$

At n=4

$$L_4 = \frac{e^4}{\varepsilon_0^2}\sqrt{\frac{G}{128\pi^3\hbar^3 c^7}}, \quad t_3 = \frac{e^4}{\varepsilon_0^2}\sqrt{\frac{G}{128\pi^3\hbar^3 c^9}}, \quad M_3 = \frac{e^4}{\varepsilon_0^2}\sqrt{\frac{1}{128\pi^3 G\hbar c}}$$

At n=5

$$L_5 = \frac{e^5}{\pi^2\hbar^2 c^4}\sqrt{\frac{G}{512\varepsilon_0^5}}, \quad t_5 = \frac{e^5}{\pi^2\hbar^2 c^5}\sqrt{\frac{G}{512\varepsilon_0^5}}, \quad M_3 = \frac{e^5}{\pi^2\hbar^2 c^2}\sqrt{\frac{1}{512 G\varepsilon_0^5}}$$

At n=0, we obtain the Planck natural units while at n=1, we obtain the Stoney units of measure

case2:

We further derive the fundamental units of measure at values of n=0, 1, 2, only for the gravitational coupling $\alpha = \dfrac{Gm^2}{\hbar c}$

<u>At n=0</u>

$$L_o = \frac{e}{m}\sqrt{\frac{\hbar}{2\varepsilon_0 c^3}}, \quad t_o = \frac{e}{m}\sqrt{\frac{\hbar}{2\varepsilon_0 c^5}}, \quad M_o = \frac{e}{Gm}\sqrt{\frac{\hbar c}{2\varepsilon_0}}$$

<u>At n=1</u>

$$L_1 = \frac{e}{c^2}\sqrt{\frac{G}{2\varepsilon_0}}, \quad t_1 = \frac{e}{c^3}\sqrt{\frac{G}{2\varepsilon_0}}, \quad M_1 = \frac{e}{\sqrt{2G\varepsilon_0}}$$

<u>At n=2</u>

$$L_2 = meG\sqrt{\frac{1}{2\varepsilon_0 c^3 \hbar}}, \quad t_2 = meG\sqrt{\frac{1}{2\varepsilon_0 c^7 \hbar}}, \quad M_2 = em\sqrt{\frac{1}{2\varepsilon_0 \hbar c}}$$

It proves difficult to deduce the Planck units here, simply because the charge and mass do not cancel out. But if you set the ratio of charge to mass at n=0 in the above formulas as $\dfrac{e}{m} = \sqrt{4\pi\varepsilon_0 G}$, one obtains the Planck units. Also, one obtains the values of n=2 in Case1 when we substitute for $m = \dfrac{e}{\sqrt{4\pi\varepsilon_0 G}}$, in Case2, for n=2. This means that, the formulas which do not exist in case2 but are present in case1 can be calculated by applying a simple formula, $e = m\sqrt{4\pi\varepsilon_0 G}$ and vise versa is true.

When we substitute for $e = m\sqrt{4\pi\varepsilon_0 G}$, at n=1, we obtain,

$$L_1 = \frac{Gm}{c^2}\sqrt{2\pi}, \quad t_1 = \frac{Gm}{c^3}\sqrt{2\pi}, \quad M_1 = m\sqrt{2\pi}$$

This represents formulae at a scale of general relativity, in which it is deduced here that, the mass of a particle in both the special and general relativity theory makes sense when multiplied by a constant $\sqrt{2\pi}$.

When $me = m^2\sqrt{4\pi\varepsilon_0 G}$ at n=2 above, we obtain

$$L_2 = m^2\sqrt{\frac{2\pi G^3}{c^5 \hbar}}, \quad t_2 = m^2\sqrt{\frac{2\pi G^3}{c^7 \hbar}}, \quad M_2 = m^2\sqrt{\frac{2\pi G}{\hbar c}} = \frac{2\pi m^2}{m_p}$$

Where m_p is the Planck mass $\sqrt{\dfrac{2\pi\hbar c}{G}}$

It should however be taken seriously from the above investigation that changing the number 2π in the formulas (case1, at n=0, the planck units/ scales), will change the statement of the formula $e = m\sqrt{4\pi\varepsilon_0 G}$, which will mean that the values of the fundamental physical constants $\dfrac{1}{4\pi\varepsilon_0}$,G are varying, therefore in order to maintain the constants unchanged we have to maintain the Planck units unchanged in formula as they are derived here. Thus the Planck mass will have a mass given by, $2.2176470119 \times 10^{-8} \sqrt{2\pi} = 5.4556 \times 10^{-8}$ Kg.

At present there is no candidate theory of everything that includes the standard model of particle physics and general relativity. For example, no candidate theory is able to calculate the mass of an electron. However in this paper the mass of an electron is deduced when n=21.32 and $\alpha=1/137.036$ as,

$$M_{21.32} = e\sqrt{\dfrac{1}{2G\varepsilon_0}\left(\dfrac{1}{137.036}\right)^{20.32}} = 9.082073363 \times 10^{-31} kg$$

$$L_{21.32} = 6.745125 \times 10^{-58} m$$

$$t_{21.32} = 2.25 \times 10^{-66} s$$

Also the proton mass is deduced at n= 18.26, and $\alpha=1/137.036$ as,

$$M_{18.26} = e\sqrt{\dfrac{1}{2G\varepsilon_0}\left(\dfrac{1}{137.036}\right)^{17.26}} = 1.688659377 \times 10^{-27} kg$$

Other masses including the mass of the earth (n= -29.99) can be deduced in the same way. It is important to note that, the value of n is negative for massive particles (e.g mass of the Sun and earth) but positive for microscopic particles like electrons and protons.

It is hereby noted that the values of the energy scales corresponding to the given length scale are off the scale and do not necessarily represent phenomenon at each given length scale. The values of these energy scales for each interaction as quoted in scientific literature will prove to be different from the ones represented here.

For example;

For atomic length scale with $l_a \sim 10^{-10} m$, the value of n to be used in calculating other scales will be given by n= -22.83, from which the energy scale can be calculated as,

$$E_{-22.83} = M_{-22.83} c^2 \sim 7.6 \times 10^{43} GeV$$

For strong interaction length scale with $l_s \sim 10^{-15} m$, the value of n to be used in calculating other scales will be given by n= -18.151, from which the energy scale can be calculated as,

$$E_{-18.151} \sim 7.6 \times 10^{38} GeV$$

For electroweak interaction length scale with $l_w \sim 10^{-18} m$, the value of n to be used in calculating other scales will be given by n= -15.343, from which the energy scale can be calculated as,

$$E_{-15.343} \sim 7.6 \times 10^{35} GeV$$

It is possible that the correct theory of everything has been found in the formulas given above. It therefore seems appropriate for the reader or researchers to consider the calculation and determination of the values of the masses of all particles at given length/time scales in the universe with explicit accuracy, even if some people may consider such an enterprise premature or foolhardy. It is worth noting that the length and time scale through which one can probe the whole mass of the earth is $4.439 \times 10^{-3} m$ and $1.4806 \times 10^{-11} s$ respectively. It is therefore important to know how one can calculate the mass of any particle with accuracy and then inquire with simplicity into the length and time scale at which such a particle can be studied. This then means that any consistent theory of nature like the one constructed would be able to deduce the required derived quantities (i.e. voltage, current, magnetic field etc) from the given formulas for fundamental physical units of measure of mass, length and time without a need to inquire into other theories like the standard model, string theory or quantum gravity.

Chapter 19. A Simpler Schrödinger Equation

Consider a relativistic particle such that its mass changes with the electromagnetic fields. Arranging Eqn (6) to get the ratio of the relative speed of the particle to that of light

$v/c = (AE\, e\, /2nhc)$

According to special relativity the ratio of the speeds are in the form

$V/c = \sqrt{(1 - m_0^2/m^2)}$

Equating the two equations and multiplying through by c^2 we get

$$E^2 - E_o^{\ 2} = \frac{m^2 c^2 k^2 e^4}{n\hbar^2} = EE_n$$

Where E is the total relativistic energy of the particle (mc^2) and E_o is it's rest energy. Multiplying both sides of the eqn by $2m/\hbar$ and taking into account that

$E_n = mk^2 e^4 / 2n\hbar^2$

We have the square of the wave length as

$$\left[\frac{\hbar}{mc}\right]^2 = \lambda^2 = \frac{\hbar^2}{2m(E - E_o)}\left[\frac{16\pi^2 E_n}{E + E_o}\right]$$

Where E_n is the energy of the quantized state of the hydrogen atom, this is the simplest form of the Schrödinger equation. When $E_o = 0$ and $E_n = E/4$ the momentum of the non relativistic particle is given by, $p = \sqrt{2mE}$. This is also called the Fermi's momentum where E is the Fermi's energy

Chapter20. The Earliest Period of Time[1] in the History of the Universe[2]

Observations have suggested that the universe began 13.7billion years ago. The universe was so hot with particles having a very high energy, in its earlier phase. The evolution then proceeded with this energy forming the first protons, electrons and neutrons, then nuclei and finally atoms. The microwave background was also emitted during the formation of the neutral hydrogen. Finally the structure of the universe was formed when matters aggregated into the first stars and quasars and on large scale clusters of galaxies and super clusters were formed.

In cosmology[3], the Planck epoch , named after Max Planck[4] Max Planck[5], is the earliest period of time[6] in the history of the universe[7], from zero to approximately 10^{-43} seconds, it is at this time that quantum effects[8] of gravity[9] were significant. At this period approximately 1.37×10^{10} years ago all fundamental forces[10] were unified. The state of the universe during the Planck epoch was unstable, tending to evolve and giving rise to the familiar manifestations of the fundamental forces through a process known as symmetry breaking[11]. It is currently believed that the Planck epoch inaugurated the Grand unification epoch[12], and that symmetry breaking quickly led to the era of cosmic inflation[13], the Inflationary epoch[14], during which the universe

1. http://en.wikipedia.org/wiki/Time

2. http://en.wikipedia.org/wiki/Universe

3.   http://en.wikipedia.org/wiki/Physical_cosmology

4.   http://en.wikipedia.org/wiki/Max_Planck

5.   http://en.wikipedia.org/wiki/Max_Planck

6.   http://en.wikipedia.org/wiki/Time

7.   http://en.wikipedia.org/wiki/Universe

8.   http://en.wikipedia.org/wiki/Quantum_mechanics

9.   http://en.wikipedia.org/wiki/Gravity

10.   http://en.wikipedia.org/wiki/Fundamental_force

11.   http://en.wikipedia.org/wiki/Symmetry_breaking

12.   http://en.wikipedia.org/wiki/Grand_unification_epoch

greatly expanded in scale over a very short period of time (see Wikipedia-Planck epoch).

The first seconds of the universe can be calculated if we apply the law of attraction to the box lowered into a black hole, we assume that there is a force acting in opposite direction to the buoyancy force (Eqn12). The forces acting on the box are different in a way that, the force expressed in equation 12 is proportional to the schwarzichild's radius while the force acting in the opposite direction is proportional to time measured in seconds. Therefore we have two different forces, one taking a particle through a distance and the other through timelines.

Therefore in time t for any particle moving through space, the force acting on such a particle will be given by,

$$F_t = \frac{e^2 c^5 \hbar}{32 \pi \varepsilon_0 G^3 m^4} = t \frac{c^7}{16 \pi G^2 m}$$

Where $t = \frac{e^2 \hbar}{2 \varepsilon_0 c^2 G m^3} = 2.54821 \times 10^{-68} \frac{1}{M^3} \ (s)$,

Thus keeping other factors constant, the time is inversely proportional to the cube of the mass of a particle. For the Planck mass of $2.1765 \times 10^{-8} kg$, the earliest period of time in the universe is $2.472 \times 10^{-45} s$. Then the total mass responsible for the current age of the universe (13.82 billion years) is, $3.880 \times 10^{-29} kg$.

13.   http://en.wikipedia.org/wiki/Cosmic_inflation

14.   http://en.wikipedia.org/wiki/Inflationary_epoch

Chapter21. New Physics: Regularization and Physics beyond the Standard Model

History tells us that if we hit upon some obstacle, even if it looks like a pure formality or just a technical complication, it should be carefully scrutinized. Nature might be telling us something, and we should find out what it is (G. t Hooft, 1997).

In physics, one of the ultimate goals is to unify the fundamental forces of nature. Today physicists have been able to unify three of the four known fundamental forces (the electromagnetic, the strong and the weak nuclear forces in a single quantum field theory-the standard model). The fourth fundamental force, gravity, on the other hand is described by the general theory of relativity. Because the other fundamental interactions are quantized, it therefore seems natural that in a grand unified theory, a theory of all the fundamental forces, gravity is quantized as well into perhaps Quantum gravity.

A theory of quantum gravity is needed to describe things that are very small but also very heavy, like black holes or the early universe. However, the development of a quantum theory of gravity seems difficult on grounds that, in general relativity all physical qualities have definite values, whereas in quantum mechanics they do not as shown in Heisenberg's uncertainty principle.

The problems in General Relativity arise from trying to deal with a universe that is zero in size (infinite densities). But quantum mechanics suggests that there may be no such thing in nature as a point in space-time, implying that space-time is always smeared out, occupying some minimum region. The minimum smeared-out volume of space-time is a profound property in any quantized theory of gravity and such an outcome lies in a widespread expectation that singularities will be resolved in a quantum theory of gravity.

However, Prof Brian Dolan at the Department of Theoretical Physics, NUI Maynooth, is quick to point out that there is not yet any set agreement on what a theory of quantum gravity should look like, or even on the exact problem it is trying to solve. "There is no accepted theory of quantum gravity," he says. "There are currently a number of contenders, and by far the most popular is superstring theory. Many physicists find superstring theory compelling due to its internal elegance, but despite decades of intense research it has not produced a single experimentally testable result." He suspects that trying to unite general relativity and quantum mechanics may be the wrong way to go, and that any future breakthrough may come from a completely unexpected direction; perhaps from some young mind with a fresh perspective.

This chapter employees new idea towards the development of a quantum theory of gravity in a bid to solve the following unsolved problems in physics;

(i) Is it true that at every spatial dimension, there exists new physics and that it is the work of Physicists to find out? What is the method or procedure through which new physics can be found? Does this require extra dimensions?

(ii) Does nature have more than four space-time dimensions? If so, what is their size? Are dimensions a fundamental property of the universe or an emergent result of other physical laws? Can we experimentally observe evidence of higher spatial dimensions?

(iv) Can the singularities that plague the General theory of Relativity be resolved in any quantum theory of Gravity?

The Standard Model is inconsistent with that of general relativity, to the point that one or both theories break down under certain conditions (for example within known spacetime singularities like the Big Bang and the centers of black holes beyond the event horizon).

The appearance of singularities in any physical theory is an indication that something is wrong and that there is a need for new physics. Singularities can be avoided in GR and any field theory through the introduction of an efficient regularization procedure as this book directs.

Regularization is a method of modifying observables which have singularities in order to make them finite by the introduction of a suitable parameter called regulator. The regulator, also known as a "cutoff", models our lack of knowledge about physics at unobserved scales (e.g. scales of small size or large energy levels). **It compensates for the possibility that "new physics" (beyond the SM) may be discovered at those scales which the present theory is unable to model,** while enabling the current theory to give accurate predictions as an "effective theory" within its intended scale of use.

The need for regularization terms in any quantum field theory of quantum gravity is a major motivation for Physics beyond the standard model. Infinities of the non-gravitational forces in QFT can be controlled via renormalization only but additional regularization and hence new physics is required uniquely for gravity. The regularizers model, and work around, the breakdown of QFT at small scales and thus show clearly the need for some other theory to come into play beyond QFT at these scales. A. Zee (Quantum Field Theory in a Nutshell, 2003) considers this to be a benefit of the regularization framework, theories can work well in their intended domains but also contain information about their own limitations and point clearly to where new physics is needed.

Therefore the main objective of this section is to discover new physics at those scales (or extra dimensions) which the General relativity theory and Quantum mechanics is unable to model. The section also sets out to prove that due to quantum gravitational effects, there is a minimum distance beyond which the force of gravity no longer continues to increase (operate) as the distance between the masses become shorter.

General Theory

During the years, strong evidence has appeared that the acceleration of any physical object cannot be arbitrarily large, but it should be superiorly limited. For example in string theory, it was derived that string acceleration must be less than some critical value, determined by the string tension and its mass. From the classical point of view (as Wheeler suggested), if we consider an extended object in **rotating motion**, we have the acceleration $a = v^2/R$ and it follows that a, must be at least limited by c^2/R. However to differ from the classical Newtonian mechanics and Einstein's General relativity theory we introduce a regulator "Cutoff" $\alpha_g{}^n$,where α_g is the gravitational coupling constant, R is the distance between two masses and n is a positive number (**extra dimension number**), then the acceleration must be limited by $a = \frac{c^2}{2R}\alpha_g{}^n$ (i), (Assuming a diameter of 2R).

Thus to avoid the infinity but while retaining the point nature of the particle would be to postulate a small additional dimension **n** over which the particle could 'spread out' rather than over 3D space.

For example, in the Unruh temperature we can only and only deduce both the Hawking temperature and maximal temperature (Sakharov Temperature) under the assumption of the existence of a maximal acceleration given in formula (i) above as,

The Unruh temperature is given as,

$$T = \frac{\hbar a}{2\pi ck}$$

Since the acceleration is known from (i) above, then the temperature will reduce to,

$$T = \frac{\hbar c}{4\pi Rk}\alpha_g^{\ n}$$

For a Schwarzschild Black hole of radius $R = \frac{2GM}{c^2}$, the temperature reduces to

$$T = \frac{\hbar c^3}{8\pi GMk}\alpha_g^{\ n}$$

Since the gravitational coupling constant has a formula $\alpha_g = \frac{GM^2}{\hbar c}$, taking values of n=0,1,2,............,N. Then the Hawking temperature will become a result of n=0 extra spatial dimensions as, $T = \frac{\hbar c^3}{8\pi GMk}$. Also the maximum temperature (Sakharov temperature) is deduced at n=1/2 as, $T = \frac{1}{8\pi k}\left(\frac{c^5 \hbar}{G}\right)^{1/2}$.

Therefore the temperature of a black hole increases as a black hole loses mass in Hawking Black hole evaporations. The analysis given above is a clear indication that the temperature doesn't increase exponentially as it has been known from Hawking's original proposals, there is a maximum temperature, a limit on temperature that screens (resolves) the classical singularity. It is therefore true that the radiation spectrum contains all Standard Model particles, which are emitted on our brane, as well as gravitons, which are also emitted into the extra dimensions. It is expected that most of the initial energy is emitted during this phase in Standard Model particles. Therefore we recommend the applications of a factor $\alpha_g^{\ n}$ in situations involving the examination and experimentation of quantum gravitational phenomenon. We shall see in the coming chapters that such a factor when used in loop quantum cosmology it reproduces both the results of loop quantum gravity and string theory.

The idea of including extra dimensions, to achieve the goal of unifying physics, is not a new one. Already the year before Einstein in 1915 introduced his theory of general relativity; Gunnar Nordstrom suggested a unification of gravity and electromagnetism with the introduction of a fifth dimension. These forces were the two only forms of interaction known at that time. But this idea was forgotten for some time with the eruption of the First World War. But in April 1919 Theodor Kaluza introduced independently, in a letter to Einstein, a fifth dimension in an attempt to unify Einstein's theory of gravity and Maxwell's theory of light. Oskar Klein (1926) contributed, in this quest, with his assumption that the extra dimension was compactified. The Kaluza-Klein theory was a fact. This theory includes an extra space dimension that is rolled up into a tiny circle, i.e. compactified. And in this five dimensional theory, there is only one underlying force, gravity. But in the four-dimensional spacetime observed at great distances, it appears to be three kinds of forces, among these a gravitational and an electromagnetic force. This topic was initially a popular topic for research, but lost much of its interest with the introduction of quantum mechanics.

In recent years the topic of extra dimensions has experienced a renewed interest. This renewed interest is also due to the exciting possibility of observing new and spectacular physical phenomena at far lower energy scales than otherwise. Even at energies available in the not so distant future, these phenomena could appear. Among these is the creation of higher dimensional semi-classical microscopic black holes. The possibility of observing these objects, is viewed as an opportunity to perhaps discover new intriguing physics.

Therefore from (i) using Einstein's equivalence principle we get the minimum distance beyond which the force of gravity no longer continues to increase as; $R = \dfrac{R_s}{\alpha_g{}^n}$ (ii). Where $R_s = \dfrac{2GM}{c^2}$ is the Schwarzschild radius.

We therefore conclude that;

(i) At n=0 extra spatial dimension, we have a physical theory of General relativity at a length scale of $R = R_s = \dfrac{2GM}{c^2}$ - the Schwarzschild radius.

(ii) At n=1/2 extra dimension, we have the quantum theory of gravity (New physics) at the Planck length scale $l_p = \sqrt{\dfrac{\hbar G}{c^3}}$.

(iii) At n=1 extra dimension, we have the theory of Quantum mechanics at the Compton wavelength scale of $\lambda = \dfrac{\hbar}{mc}$.

(iv) Lastly at n=2 we have new physics at a length scale $R = \dfrac{h^2}{GMm^2}$ and the journey continues.

According to the Standard Model of particle physics, the world is governed by four fundamental forces: gravity, electromagnetism, and the weak and strong nuclear forces. Although things act a bit "spooky" down on the quantum level, science has managed to generally describe all of these forces at both the macro and quantum scales – except gravity.

Gravity is the weakest of the fundamental forces, and it's been suggested that this is because some gravitons (the hypothetical particles) that carry the gravitational force tend to escape into extra dimensions. We're simply too big to travel through or even notice these other dimensions.

So, to study whether these extra dimensions are lurking in extremely tiny spaces, the researchers from Osaka, Kyushu and Nagoya Universities set out to test gravity on the sub nanometer scale. To do so, they used the world's highest intensity neutron beam, which is housed at the Japan Proton Accelerator Research Complex (J-PARC).

The team found that the results matched predictions based on the known laws of physics, which indicates that Newton's law still applies as expected down to a scale of less than 0.1 nanometers. No unexplained force ie, another dimension is acting on these particles at this scale.

That doesn't mean those extra dimensions aren't there, just that they may be hiding at even smaller scales still. The researchers are currently working to further improve the sensitivity of the equipment, which might help them probe those tiny spaces.

In a completely different context, an international team of researchers led by Professor Immanuel Bloch (LMU/MPQ) and Professor Oded Zilberberg (ETH Zürich) has now demonstrated a way to observe physical phenomena proposed to exist in higher-dimensional systems in analogous real-world experiments. Using ultracold atoms trapped in a periodically modulated two-dimensional superlattice potential, the scientists could observe a dynamical version of a novel type of quantum Hall effect that is predicted to occur in four-dimensional systems. (Nature, 4 January 2018)

"Physically, we don't have a 4D spatial system, but we can access 4D quantum Hall physics using this lower-dimensional system because the higher-dimensional system is coded in the complexity of the structure," a researcher with the US-based team, Mikael Rechtsman from Penn State University, told Ryan F. Mandelbaum at Gizmodo. "Maybe we can come up with new physics in the higher dimension and then design devices that take advantage the higher-dimensional physics in lower dimensions."

The above statements can be summed up in the following simplest model;

Let the Gravitational force between two identical particles be related to the magnetic force between them and similarly let the electric force between two particles be related to the magnetic force as;

Gravitational force $(\frac{Gm^2}{R^2})$ = magnetic force (Bec) $\times \alpha_g^{\ n}$ (iii)

and

Electric force $(\frac{e^2}{4\pi\varepsilon R^2})$ = magnetic force (Bec) $\times \alpha_e^{\ n}$ (iv)

Where α_e is the electromagnetic coupling constant- Fine structure constant

The magnetic flux, represented by the symbol Φ, threading some contour or loop is defined as the magnetic field $\mathbf{B}$ multiplied by the loop area, $A = \pi \mathbf{R}^2$, i.e. $\Phi = \mathbf{B} \cdot \mathbf{A}$. Obviously, both $\mathbf{B}$ and $\mathbf{A}$ can be arbitrary and so is Φ. The inverse of the flux quantum, $1/\Phi_0$, is called the **Josephson constant**, and is denoted K_J.

However, if one deals with the superconducting loop or a hole in a bulk superconductor, it turns out that the magnetic flux threading such a hole/loop is quantized. Therefore the magnetic flux quantum from (iii) and (iv) will be given by $\Phi_G = \pi G m^2 / ec\, \alpha_g^{\ n}$ and $\Phi_E = e/4\varepsilon c\, \alpha_e^{\ n}$ respectively. Such that at n=0 extra dimension, $\Phi_G = \pi G m^2 / ec$ and $\Phi_E = e/4\varepsilon c$ representing the classical flux at 3D spatial dimensions.

At n=1/2 extra dimension, $\Phi_G = \frac{\pi m}{e}\left(\frac{Gh}{c}\right)^{1/2}$ and $\Phi_E = \left(\frac{\pi h}{4\varepsilon c}\right)^{1/2}$ representing the quantum theory of Gravity.

At n=1 extra dimension, $\Phi_G = \pi h/e$ and $\Phi_E = \pi h/e$ representing the magnetic flux quantum at the quantum scale. Also at n=1 the magnetic flux value is the same in both equations, meaning that the gravitational force becomes analogous to the electromagnetic force at n=1.

In other words, just as a 3D object casts a 2D shadow, scientists have managed to observe a 3D shadow potentially cast by a 4D object – even if we can't actually see the 4D object itself. That could unlock some new findings in the very fundamentals of science.

Chapter22: A Grand Unification

Physicists have argued out that the more elegant and symmetrical the theory is, the more it is beautiful. The elegancy of any physical theory is suspected at a level to which it holds well with other theories , that is ,the capability of the theory to conform with the well known laws of nature at all levels.

In this section we examine the mechanism through which quantum mechanics becomes comparable with gravity and the scale at which this occurs. At the Planck scale all interactions (the weak interaction, strong interaction and electromagnetism) are assumed to merge into a single interaction that alone occurs at very high energies of about 1 TeV. The equations that do describe this phenomenon are not yet found and therefore requires one's deep effort to capture the reality of this entire puzzle.

To capture interest in these interactions we need to know first, their strength and second the range in which they occur. The strength defines the coupling constants and the range defines the attractions, on the other hand the coupling constant determines the strength of any interaction and therefore is a number in a sense that it is a dimensionless constant. A coupling constant is a very important quantity in dynamics, for example, in the motion of a large lump of magnetized iron, the magnetic forces are more important than the gravitational forces because of the relative magnitudes of the coupling constants.

The standard model is a theory of three fundamental forces - electromagnetism, weak interactions and strong interactions; however, these three forces are not tied together Howard Georgi and Sheldon Glashow discovered that the Standard Model particles can arise from a single interaction, known as a grand unified theory. Grand unified theories predict relationships between otherwise unrelated constants of nature in the Standard Model. Gauge coupling unification is the prediction from grand unified theories for the relative strengths of the electromagnetic, weak and strong forces and this prediction was verified at LEP in 1991 for supersymmetric theories.

In particle physics, supersymmetry (often abbreviated SUSY) is a novel symmetry that relates elementary particles of one spin to another particle that differs by half a unit of spin and are known as superpartners. Since the particles of the Standard Model do not have this property, supersymmetry must be a broken symmetry allowing the 'sparticles to be heavy.

One of the main motivations for SUSY comes from the quadratically divergent contributions to the Higgs mass squared. The quantum mechanical interactions of the Higgs boson causes a large renormalization of the Higgs mass and unless there is an accidental cancellation, the natural size of the Higgs mass is the highest scale possible. This problem is known as the hierarchy problem Supersymmetry reduces the size of the quantum corrections by having automatic cancellations between fermionic and bosonic Higgs interactions. If supersymmetry is restored at the weak scale, then the Higgs mass is related to supersymmetry breaking which can be induced from small non-perturbative effects explaining the vastly different scales in the weak interactions and gravitational interactions. The failure of experiments to discover either supersymmetric partners or extra spatial dimensions, as of 2006 has encouraged loop quantum gravity researchers.

The determination of the strength of the forces

We assume a model that explains everything on the length scales, the best scale so far we are familiar with is the Planck length scale, however in this model we don't associate ourselves in knowing this scale and therefore develop new scales that alone are combined together to lead to some observable phenomenon describing the forces involved in the interactions. The equation describing the model is developed and given by;

$$(v^2/c^2 + n^2 \beta_{Qo}) = 8\pi \beta_{gEo} \ (1)$$

Where β_{Qo} is a length ratio given by l_Q/l_o, in this case $l_Q = hc/W$, $\hbar$ is Dirac constant, c is the speed of light and W is the energy. Also $\beta_{gEo} = l_{gE}/l_o$ where $l_{gE} = 8\pi G M_{gE}^2/W$, G is the universal gravitational constant , M_{gE} is the mass of a particle in the combined fields given by P_{gE}/c where $P_{gE} = Gm^2 ke^2/R^2 c^2$,is the momentum for an elementary particle of mass m and an elementary charge e , k is the coulomb constant and R is the distance between any two particles. The equation here addresses the problems in form of length scales simply because it is at these scales that quantum mechanics seem to be comparable to gravity. The momentum P_{gE} is a momentum of a particle experiencing the strength of the electromagnetic fields and gravity.

The strength is determined by a very small coupling constant as we shall see later. The smaller the distance between elementary particles, the higher the momentum and vice versa is true.

The exchange of photons between an electron and a proton in an atom is explained by Quantum Electrodynamics (QED), with a coupling constant determining the strength of the electromagnetic force. The equation of the interaction responsible for QED on the length scale, which is the Compton length, is given by the equation

$$\Sigma \psi^2 \text{fiti} = 2\pi \beta \text{RCE} \text{ (2)}$$

The expression $\Sigma \psi^2$fiti is the force changer where $\psi_{fi} = fiR^2/ke^2$ and $t=\{fi^2ke^2/R^2\}/Fn^3$,

βRCE ,remains a constant given by lc / lRE (lc is the Compton length $\hbar/mc$ and lRE=ke^2/mR^2c^2).

On multiplying both sides of Eqn1 by a quantity $\Sigma \psi^2$fiti we obtain,

$$(v^2/c^2 + n^2 \beta Qo) \Sigma \psi^2 \text{fiti} = 8\pi \beta \text{gEo} \Sigma \psi^2 \text{fiti}$$

We then examine the condition for which βQo will be a maximum and minimum. It is found out from relativity that βQo is maximum when the lorentz factor $\gamma = (1-v^2/c^2)^{-1/2}$ is very small that is,

$\gamma = 1/n\sqrt{\beta Qo}$ **or when the velocity v = c** $\sqrt{(\Sigma \psi^2 \text{fiti} - n^2 \beta QO)}$

We hence obtain a general interaction equation as,

$\Sigma fy^3 \psi^2$fiti$=2\pi\beta$RCEΣFn^3 /$\xi\beta$gEo , n= 1,2,3 (3)

The following conditions are then taken into account

1) For lo = lx = mc^2/Fp, βgEo=βgEx =lgE/lx .

2) For lo = lc , βgEo=βgEQ =lgE/lc , and

3) For lo = ls = Gm/c^2, βgEo=βgEs =lgE/ls , which gives

$\Sigma fy^3 \psi^2$fiti$= F1^3 + F2^3 + F3^3 = 2\pi\beta$RCE (Fp^3 /$8\pi\beta$gEx + Fp^3 / $256\pi^3\beta$gEQ+ 2FB^3 / $\pi \beta$gEo) (4)

Where $Fp = c^4/G$ is the Planck unit force and $FB^3 = m^2c^3/\hbar$ is the force required for strong and weak interactions to take place.

Again setting a condition,

For for $l_o = l_z = ke^2/mc^2$, $\beta_{gEo}=\beta_{gEz} =l_{gE}/l_z$.

$\Sigma f_y^3 \psi^2 f_i t_i = F_4^3 = 2\pi\beta_{RCE} (F_z^3/32\pi^3\beta_{gEx})$

Where $F_z^3 = m^2c^4/ke^2$,

Also for $l_o = l_N = \hbar^2 m^3 G^2/k^3 e^6$, $\beta_{gEo}=\beta_{gEN} =l_{gE}/l_N$, we obtain,

$\Sigma f_y^3 \psi^2 f_i t_i = F_5^3 = 2\pi\beta_{RCE} (F_z^3/2\pi^3\beta_{gEN})$ (5)

Measuring the value of the strong, weak and electromagnetic coupling constants gives us away through which we can determine supersymmetric levels. From supersymmetry and grand unification of elementary particles the couplings agree to 1%. The relationships of the sum of the cubes of the forces to each individual cube of the force, and that of the sum of the square of masses with each known mass squared casts much information about the masses and couplings of the supersymmetric particles as shown below, when Eqn4 is divided through respectively by the cubes of the forces F_1^3, F_2^3 and F_3^3 the following equations are obtained,

$\Sigma F_n^3/F_1^3 = 1+16\alpha_g^3+1/32\pi^2\alpha_g$ (6)

$\Sigma F_n^3/F_2^3 = 1+32\pi^2\alpha_g+512\pi^2\alpha_g^4$ (7)

$\Sigma F_n^3/F_3^3 = 1+1/6\alpha_g^3+1/512\pi^2\alpha_g^4$ (8)

$\Sigma F_n^3/F_4^3 = 1+\beta^2(4\pi^2+1/8\alpha_g)+64\pi^2\alpha_s^2\alpha_g$ (9)

Where $\beta = ke^2/Gm^2$ is the ratio of the fine structure constant α_s to the gravitational coupling constant α_g, given respectively as $\alpha_g = Gm^2/\hbar c$ and $\alpha_s =ke^2/\hbar c$.

Now equating $F_4 = F_5$, $F_5= F_3$, $F_5= F_1$ we obtain; m_1, m_2, $m_3$and $m_4$respectively, Adding the squares of the masses we obtain,

$\Sigma m_n^2 = m_1^2+m_2^2+m_3^2+m_4^2$ (10)

Which gives the sum per unit mass as,

$\Sigma m_n^2/m_1^2 = 1+ 16\pi^2\alpha_s^4 + (8\pi/\alpha_s)^{1/2} + 4/(128\alpha_s^4)^{1/5}$ (11)

$\Sigma m_n^2/m_2^2 = 1+ 1/16\pi^2\alpha_s^4 + (1/8\pi\alpha_s^9)^{1/2} + (1/4\pi^2)(128\alpha_s^{24})$ (12)

The equations generated so far give a basis for the nature and type of supersymmetry exhibited by a particle experiencing forces at both the Planck and grand unified scales. It is thus shown here that the electromagnetic coupling constant is a result of mathematically summing the squares of the masses generated and then dividing through by the square of the mass in the summation while the gravitational coupling constant is the result of summing the cubes of the forces and then dividing through by the cube of the force in the sum. This idea at its best is taken to be the basis for symmetric theories as we shall see in the results obtained.

Results

(1)The unification of coupling calculations

At equal forces that is $F_1 = F_2 = F_3 = F_p$ the mass $M_p = (\hbar c / 8\pi G)^{1/2} = 2.1765 \times 10^{-8}$ kg, is obtained which is the Planck mass for which the Schwarzschild radius is equal to the Compton length divided by π^1.When Eq4 is divided through by F_1^6 and F_2^6 we obtain equations of the form;

$$\Sigma F_n^3 / F_1^6 = \Omega / F_p^3 \ (13)$$

$$\Sigma F_n^3 / F_2^6 = \text{\euro} / F_p^3 (14)$$

Where, $\Omega = 4m^2 / m_p^2 + 1/\pi + m^8 / 32\pi^2 m_p^8$,

$\text{\euro} = m^6 / 8\pi m_p^6 + 16\pi m^4 / m_p^4 + 2m^{12} / \pi m_p^{12}$

The mass relations equations obtained above indicate the scale at which gravity may be strong and weak. Obtaining these results on the Planck force and mass scale is evidence for the existence of the theory of quantum gravity. The values Ω and $\text{\euro}$ represent a series equation defined by increasing powers in the mass ratio (m/m_p). The mass m is assigned to any particle and the mass m_p is assigned to the Planck scale defining quantum gravity.

The unit of energy is $M_p c^2$; the unit of electric charge is $\sqrt{hc/k}$, where k is coulomb constant and so forth. On the other hand, one cannot form a pure number from these three physical constants. Thus one might hope that in a physical theory where $\hbar$, c, and G were all profoundly incorporated, all physical quantities could be expressed in natural units as pure numbers. Within its domain, this paper has achieved it for example, imagining that there were

---

just two quark species with vanishing masses. Then from the two integers 3 (colors) and 2 (flavors), $\hbar$, and c (without mass parameters), the spectrum of hadrons with mass ratios and other properties close to those observed in reality, emerges by through calculation (Ω and €) as indicated from Eqn13 and Eq14 shown above. The overall unit of mass is indeterminate, but this ambiguity has no significance within the theory itself. The results obtained show an ideal Planckian theory that alone does not contain any pure numbers as parameters.

Thus, for example, the value $m_e/m_p = 10^{-22}$ of the electron mass in Planck units is obtained from a dynamical calculation. This ideal might be overly ambitious, yet it seems reasonable to hope that significant constraints among physical observables will emerge from the inner requirements of a quantum theory which consistently incorporates gravity. The model therefore provides; first, the unification of couplings calculation. second, it points to a symmetry breaking scale remarkably close to the Planck scale (though apparently smaller by 10^{-2} to 10^{-3}), so there are pure numbers with much more 'reasonable' values than 10^{-22} to shoot for. Third, it shows quite concretely how very large scale factors can be controlled by modest ratios of coupling strength, due to the logarithmic nature of the running of couplings (so that 10^{-22} may not be so 'unreasonable' after all).

While the above result is based on the study of the strength of the gravitational force, we now look for ways in which we can examine the strength of the electromagnetic force depending on the mass. This is done by dividing the sum of the squares of the masses (Eqn10) by the fourth power of the individual masses hence,

$\sum m_n^2/m_2^4 = \omega/m_G^2$ (15)

$\sum m_n^2/m_E^4 = \lambda/m_p^2$. (16)

Where $\omega = 1/16\pi^4 \alpha_s^9 + 1/4\pi^2 \alpha_s^5 + 1/(512\pi^8 \alpha_s^{19})^{1/2}$

$\lambda = 128\pi^3 \alpha_s^6 + 8192\pi^5 \alpha_s^{10} + 128\sqrt{\pi^3} \alpha_s^{11} + 128(\pi^{15} \alpha_s^{26})^{1/5}$

$m_E[(1/8\pi Ke^2)(\hbar^3 c^3/G)^{1/2}]$ is the mass obtained when $F_4^3 = F_3^3$, and

$M_G = (Ke^2/G)^{1/2}$ is the mass obtained when the electromagnetic force is equal to the gravitational force.

It can now be theorized that the strength of the electromagnetic force is determined by Eqn15 and 16 at which a series power equation in the fine structure constant defined by ω and λ is a constant.

(2) The length scales at which the masses predicted by the standard model survive

The mass of the W and Z bosons (M_W, M_Z), Higgs particle (M_H) and the mass scale at the grand unification (M_{GUT}) are generated. We multiply a coupling constant μ with the force F_3^3, of which we equate to F_4^3 that is;

$$\mu\, F_3^3 = F_4^3$$

From which

$$\mu = R_B^2 / R_o^2,$$

Where R_o is the length scale determined experimentally and

$R_B = (8\,\pi G k e^2/c^4)^{1/2} = 6.9101 \times 10^{-36}$ m, which is greater than the Planck length.

So the equation that produces the different masses at R_o will be given by the square of the mass as,

$$M^2 = m_p^2 / 8\,\pi\mu\, \alpha_s^2$$

Where $\alpha_s = 1/137$, is the electromagnetic coupling constant.

To obtain the masses, we need to find the length R_o , theoretically we develop the lengths given by; 1.03741×10^{-39} m, 8.3182×10^{-54} m, 9.4334×10^{-54} m, and 1.2345×10^{-53} m.

Following the given lengths we respectively obtain the masses;

$M_{GUT} = 10^{16}$ GeV, $M_W = 80.18$ GeV , $M_Z = 90.82$ GeV, and $M_H = 119$ GeV respectively.

But at $R_B = R_o$, the mass $M_B = 6.661 \times 10^{19}$ GeV is obtained. And at $R_o = 2.529 \times 10^{-37}$ m, the Planck mass is obtained (that is $M = m_p$). Therefore it is found out that the W and Z boson particles survive in length of 10^{-54} m . The Higgs particle survives to a length greater than that of the W boson $\geq 10^{-53}$ m. And finally particles at the grand unified scale will survive at 10^{-39} m.

(3) The big bang acceleration and proton decay

For proton decay the intensity P is used such that at Schwarzschild radius R and Planck mass scale m_p the life time of the proton as explained by SUSY is seen to agree so well with the

$T(time) = \alpha^2 \, m^5_p \, R \, / \, 4096 \, \pi^3 mk^4 \, \hbar$

such that at $mk = 7.96 \times 10^{-29}$ GeV, $T = 10^{35}$ yrs.

We have obtained the lifetime of protons and the mass of a particle produced during the decay process. The mass of the particle obtained is very small and can therefore be taken to be a neutrino.

The force F_3 can be expressed in the form,

$F_3 = a_3(m_3^5 \, / \, 16\pi^2 m_p^2)^{1/3}$

Where a_3 is the acceleration, this acceleration at a Planck scale will be given by

$a_3 = (c^{11} \, / \, \hbar \, G^2 m)^{1/3} = 2.4772 \times 10^{52} m/s^2$

This is quite a very large acceleration and therefore defined as the acceleration of particles during the early formation of the universe.

The results obtained describe super symmetry which is a theory required for the unification of everything we know about the physical world into a theory of everything. Significantly a larger enterprise of the theory is to produce a theory of quantum gravity which is required for the unification of general relativity with the standard model, which explains the other three basic forces in physics (electromagnetism, the strong interaction, and the weak interaction), and provides a palette of fundamental particles upon which all four forces act. Theoretically the results obtained (Eqn11 and Eqn12) show a huge correction to the particles' masses, which without fine-tuning will make them much larger than they are in nature. The problem of the unification of the weak interactions, the strong interactions and electromagnetism is solved mathematically, through the comparisons of the cube of the forces in a ratio that generates the gravitational coupling constant power equation.

The Planck mass is the mass of a black hole whose Schwarzschild radius multiplied by π equals its Compton wavelength. The radius of such a black hole is roughly the Planck length, which is believed to be the length scale at which both general relativity and quantum mechanics simultaneously become important. In accordance with the results obtained it is seen that the Planck mass is the mass at which the four forces (F_1 , F_2 ,F_3 and F_P) are equal, the forces are then taken to be related to the origin of the universe simply because at those high energies that formed the dense soup of the universe the forces were equal and the masses probing the Planck mass scale that is black holes were produced, hence those four forces a significant in that they play a crucial role in the formation of black holes. The intensity P on the other hand explains a phenomenon that occurs at the cosmic scale, for example it explains the nature of Black holes and the age of the universe. The acceleration obtained is so large that it is the acceleration that the universe had at the instant after the big bang. Obtaining this acceleration is the possibility of studying the rate of expansion of the universe at large, the accelerating universe is therefore the observation that the universe appears to be expanding at an accelerated rate.

At the Planck scale the descriptions of subatomic particle interactions in terms of quantum field theory breaks down. Also at the same scale, the strength of gravity is expected to become comparable to the other forces, mathematically all the fundamental forces are unified at that scale. The results obtained explain both the weak and strong interactions that at a length between 10^{-37}m and10^{-35} the Planck scale is attained also at lengths10^{-39}m , the grand unified scale becomes relevant , but for lengths10^{-53}m and 10^{-54}m, the standard model holds on well. We have therefore attained a unification that increases from about 10^{-59}m (standard model) to10^{-35}m (quantum gravity).The paper there fore gives out the relationship between elementary particle physics and astrophysics at a large scale.

Basing on the results obtained, it is now clearly justified that gravity can be integrated with quantum mechanics at the Planck scale. And therefore the success of the "standard model" which includes both the electroweak theory and quantum chromodynamics can now be regarded as successful in providing accurate descriptions of the fundamental particles and their interactions.

Chapter23: Is It Possible That There Is A Universe In Every Particle?

By definition an atom is the smallest unit of an element that retains the chemical properties of that element. An atom has an electron cloud consisting of negatively charged electrons surrounding a dense nucleus.

For so long scientists have been extracting electrons from atoms and bringing them onto the earth. Remember, atoms have been into existence onto the earth for many years. To me they seem like small universes within a big universe and a combination of them forms the universe within which we live. There can be ways for us to travel to those atoms and see the life there. You think this is fiction, no you are wrong. Look if the atoms merge to form the universe then the universe will separate into small parts to form the atom and this will reduce our size and shape such that we have the dimensions of the universe to which we can survive as you see it today.

If these electrons have the ability to live in the universe then also there must be a possibility for man to live on an atom. For example consider a person going to stay on an atom, remember the atom is very small and man can't even stand on it that is what you think. You will also think that he can't even see where he will land because he can't see the atom with his naked eyes. But the solution is here for a person to stay on the atom.

(i)The first ideological formulation of the possibility is by assuming that we are small towards an atom and an electron is big towards the planet.

(ii)The second assumption is that, gravity exists on an atom

(iii)The third assumption is that the space surrounding the atom is in a frozen state , that is solid like and man can move on it without floating into space

(iv) The fourth is that the life on the atom for both plant and man is constant

From the above assumptions, you can now imagine bringing a small particle like an electron from the atom onto the earth. Of course it would be easier when you take the above assumptions to be true.

(i)The electron it's self will assume to be big for it to stay on the earth and so it does

(ii) It will also assume that the space is in a frozen state for it to move on it

(ii) It will also assume that life is constant and extra

The above assumptions are true only if you set yourself into motion to the atom. The atom may be an interesting place for you than our universe. **But where is the way to it?**

The way to the atom is your imagination or simply what you think. See if you imagine that you are a small particle then that what you are but when you think that you are a big particle then the impression is true that you are. To go to the atom takes a few seconds than going to the moon. There is of course no transport but distraction of the body.

Your body will break into small particles where by each will be able to land on the atom and finally they will merge at exactly the same dimension as the atom that will make your size and shape fill comfortable on it.

Chapter24: The Principle of Least Action: An Exceptionally Simple Theory of Quantum Gravity

It seems as though we must use sometimes the one theory and sometimes the other, while at times we may use either. We are faced with a new kind of difficulty. We have two contradictory pictures of reality; separately neither of them fully explains the phenomena of light, but together they do.

Albert Einstein

In this section a new approach towards Quantum Gravity is presented, we try to study the theory from new assumptions which are far more different from models that have been used by scientists for centuries. Most physicists have clung to old models or complex mathematical scientific methods to explain phenomenon. They are trying to explain physics using the mathematics that was earlier used by Einstein, Richard P. Feyman etc. The mathematical ideas that were presented by these physicists were complicated and such has been difficult to understand and of course misleading (Read Lost in Math, a book by Sabine Hossenfielder). For example, in a statement by Dr Lee Smolin, "the mathematisation of physics has resulted in the reduction of the cosmos to a mathematical entity, which has not only confused physicists but accounts for their worst and most distracting assertions".

There is a wide spread speculation that the mathematical formulation of physics has not only confused physicists but has also lead to failures in the development of a quantum theory of gravity.

Physics as a subject should be simple and elegant, trying to explain everything from one source. In other words trying to explain all of physics from one equation call it "**the principle of least action**". Imagine deducing the equations of gravity, quantum mechanics, electromagnetism, heat etc from one equation, wouldn't it be unique than holding about ten books about a different subject of physics each starting from its own source?

The principle of least action in simple terms means; to understand how to get from point a to point b using the least amount of physical work for example taking an elevator rather than using the stairs. In other words deducing the most fundamental physical equations from one principle as described below,

Let there be a Black hole emitting particles at any given wave length and moving at a speed of light. Let these particles (say an electron) behave as a particle and at the same time as a wave. Such that, the quantum mechanic momentum is $P1= hf1/c$ where h is Planck's constant, f1 frequency of the wave (quantum wave) and c speed of light.

Then the classical electromagnetic momentum of an electron will be given as, $P2=Ee/f2$ (E electric field and e is charge on an electron) f2 is the frequency of the electron.

Then the wave-particle duality of the emitted electron will be given as the product of the two momenta as,

$P1P2 = (hf1)(Ee) / (cf2)$

But with electromagnetic waves $E= Bc$ (where B is the magnetic field of an electron) and on condition that, if f1=f2 also $P1P2= P^2$ hence the common momentum will be given by,

$P=(Beh)^{1/2}$

Now let the position of an electron in space be r, then the angular momentum of a particle Pr is quantized in units nℏ which gives,

$Pr= (Beh)^{1/2}r = nℏ. x$

The above equation is therefore proof for the existence of a *quantized electromagnetic field the kind described in Quantum Electrodynamics (QED)*

Below we show how this equation can be used in studying the properties of a Black hole the kind which were earlier deduced by Stephen Hawking. We also deduce the Stefan radiation law and the thermal equation for solids.

Quantizing Gravity (QG)

Let the position of the electron be the Schwarz child's radius $r= GM/ c^2$, Where M is the mass of the Black hole. We have assumed that the electron is at the surface of the black hole (Event horizon). Then from equation x, on arranging we have,

$[8\pi G/ c^4][Ee] [GM^2/c] = n^2ℏ$ 1

Where G– gravitational constant, n- quantum number

As you can visualize, Ee is the electromagnetic force, and $8\pi G/\,c^4$ is the gravitational force at the swcharzichild's radius. This implies that, both the electromagnetic and gravitational fields are quantized into the quantum-gravitomagnetic field.

Applications Of Equation 1
(1)To Hawking radiatons
The temperature of a black hole
On arranging equation one to get the random translational kinetic energy, we obtain

$[GM/ c^2]Ee = n^2 c^3 \hbar /8\pi GM =kT$

Where k is the boltzmann's constant and T is the temperature. Hence at n=1,

$T = c^3 \hbar /8\pi GMk 2$

Time taken by a black hole to evaporate
On dividing through eqn1 by the momentum Mc we obtain, the time t given by

$t= Mc/Ee = 8\pi G^2M^3/n^2 \hbar c^4$

Such that when n=0.03953

$t= 5120\pi G^2mo^3/ \hbar c^4 3$

Entropy of a black hole
Squaring both sides of equation 1 and arranging we generate the intensity as

$W/tA = E^2e^2/2nh = n^3 c^{10} \hbar /256\pi^3 G^4M^4 4$

Where A is the area on which the radiations fall, W is energy, and t is time

But entropy is energy divided by temperature Eq so then

$W/T = S = (n^3 c^{10} \hbar /256\pi^3 G^4M^4)(tA/T)$

Since t is known from Eq3 and T from Eq2 then at n= π

$S = Akc^3 /4G \hbar$

(2)Thermal properties of solids (wiedemann fanz law
From the intensity equation4,

$E^2e^2/2nh = c^{10} \hbar /256\pi^3 G^4M^4$

Arranging the above equation to introduce in the translational kinetic energy obtained above $[kT= c^3 \hbar /8\pi GM]$, we have

$\pi M^2G^2E^2/3c^4 =(\pi^2/3e^2)(c^3 \hbar /8\pi GM)^2=(\pi^2/3e^2)T^2k^2$

Dividing both sides by T we obtain on the left hand side of the equation the ratio of thermal conductivity K to electric conductivity δ as

$K/\delta = (\pi^2/3)(k/e)^2 T$

(3) Stefan's Radiation Law

Still from the intensity Eqn4 we can arrange the expreession on the left hand side of the equation, to read as

$$W/tA = E^2e^2/2nh = (1.875n^3/\pi^2)(\pi^2/60h^3c^2)(c^3\hbar/8\pi GM)^4$$

This is the same as Eqn4 only that it is arranged to predict something

But n=1 and $(c^3\hbar/8\pi GM) = kT$, hence the rate at which energy is radiated is given by

$$W/t = A(1.875/\pi^2)(\pi^2/60h^3c^2)(Tk)^4 = 0.19\sigma AT^4$$

Where $\sigma = \pi^2/60h^3c^2$ is the Stefan boltzmann's constant

This is Stefan's radiation law.

In conclusion, I encourage further research into this field. In other words this could be a stepping stone towards the development of a theory of everything via a simpler path of least action.

Summary of the above given approach (Gravitational Waves)

The development of a unified field theory of physics over so many years has involved the use of complicated mathematics coupled with assumptions and imaginations that have made it impossible for the creation of a simple and beautiful unified field equation. The mathematics for unification should be concise and elegant.

Attempts to unify gravitation and electromagnetism came a long way with the realm of Albert Einstein in what was called a classical unified field theory. However such attempts were abandoned with the development of modern theories of physics i.e. The General Relativity theory and the Quantum theory. These two theories are elegant and predictive in their domain of applicability.

A classical unification theory should reproduce these two theories and also be able to explain it. Today we know of the standard model and general relativity and how they are successful, but the unification of the two theories has also proved to be difficult. In my conscience these should be fruits of a classical unified field theory. In other words a classical unification incorporates them as one that is they are supposed to appear in the unification equation. Therefore the development of modern physics acts as a check for classical physics in any field of physics concerned with unification.

The objective of this paper is to develop a unified field theory uniting gravity and electromagnetism using classical physics and non- quantum approaches.

The unified electromagnetic gravitational oscillations

The gravitational potential field (gravity) is caused by the presence of any mass in space. Like gravity, charge is a fundamental property of the universe. Just as mass causes a gravity field, charge causes an electromagnetic field. There are no known methods of insulation of gravitational field in modern science. It is impossible to imagine space and gravitation separately. Gravitation exists everywhere where there is some space. The gravitational field created by all masses of our metagalaxy is the aether in which cosmic objects moves and electromagnetic oscillations are propagated. The space surrounds us since the whole matter carries gravitational charge of only one sign.

A dipole antenna is used to produce overlapping electric and magnetic fields E and B respectively. Such an antenna is so far the simplest practical antenna from a theoretical point of view, and therefore is the basis on which this study is conducted. The current amplitude I for such an antenna decreases uniformly from maximum at the center to zero at the ends.

The work done to move the electron at a speed $v(t)$ up and down the antenna is the gravitational potential energy W_G and is proportional to the velocity $v(t)$. The electric potential energy W_E of the electron at a distance R from the antenna is proportional to their angular momentum L. $W_G = k_1 v(t)$ and $W_E = k_2 L$, if the proportionalities are true then also the following should be true $W_G W_E = \beta L(t) v(t)$ where $k_1 k_2 = K^2 = \beta$ is the force required to accelerate the electrons.

Differentiating W_G with respect to height h moved by a charge up the antenna, and W_E with respect to the charge q on the electron i.e. $dW_G/dh = mg$, $dW_E/dq = ER$. Then the product of the acceleration due to gravity g and the electric field E is

$Eg = (\beta L/mRt)$div 1

$\nabla = $ div $= d/dq$

When a charge is moved up and down the antenna by the oscillator in the y direction a changing magnetic field is produced in the z direction leading to a changing gravitational field. Then the acceleration due to gravity is of magnitude $g= vc\Omega$, Ω is the curvature of space in regions near the antenna where a charge experiences the electric and magnetic field. And c is the speed of the electromagnetic wave transmitted from the antenna. Substituting for g in equation 1 gives

$\nabla \beta = d\beta/dr = f_e \Phi$ 2

This is the rate of change of force with the radius of a spherical wave front $r=cT$, T is the time- dimension and then $r^2 = x^2 + y^2 + z^2 + T^2$. Therefore $\nabla \beta$ depends on the electromagnetic force $f_e=Eq$ and the curvature $\Phi=\sqrt{BI/W}$. where W is the kinetic energy of the electron and I is the current.

According to R.L. Collins, (1997-2006) Energy is exchanged between neutral masses, via a long range electromagnetic force, and that this exchange of energy reproduces the effects of gravity. If β is taken as the gravitational force acting perpendicular through the center of mass of the particle then the potential gravitational gradient will be given by

$$\nabla^2 W_G = f_e \Phi \quad 3$$

The electromagnetic wave transmitted from the antenna will move along the x-axis. Since this wave is moving through a vaccum then there can be no conventional currents, but there is a displacement current, hence I =0 and Φ=0. Then the condition $\nabla^2 W_G$ =0 is fulfilled. The accelerated motion of electrons up and down a straight-rod dipole antenna therefore produces electromagnetic waves that satisfy the above equation. The electric, magnetic and gravitational fields are periodic in both time and space.

Change of both electric and gravitational field results in the creation of a magnetic field in the region of space-time which has a dual electrogravitational nature. The amplitude of electromagnetic and magnetogravitational constituents of the unified electromagnetic gravitational oscillations depends on field potential of opposite nature. The electromagnetic constituent is determined by gravitational potential and the magnetogravitational one is determined by electric potential. Transference of gravitational masses of matter in electrogravitational field-aether causes the creation of the proper magnetic field. (V.Ya.Kosyev, 2000).In the quantum realm, the gravitational force is so weak that it is difficult to observe quantum effects caused by gravity. However Nesvizhevsky and collaboratorshave reported an experiment in which they observed quantum effects of gravity on the behaviour of ultracold neutrons (UCNs). These neutrons have kinetic energies so low that they can be trapped by gravity above a reflecting surface. (Thomas J. Bowles, 2002).

The true displacement current: Was Maxwell Wrong?

We can generate the new displacement current which appear in Maxwell's electromagnetic theory.

Using the new physical law that the kinetic energy W of the electric charge q in a vacuum in which both the magnetic field and gravitational field exist is proportional to the acceleration due to gravity g given by,

$$W = \mu q^2 g / 4\pi$$

Where μ is the permeability of free space and g is the acceleration due to gravity,

This kinetic energy is the same as that which appear in Eqn3 $\nabla^2 W_G = f_e \Phi$ hence the current in space where no charge appears is given by

$I = \mu g \, [\nabla^2 W_G]^2 / 4\pi E^2 B$

Thus this equation shows how the current varies with g, the electric field E and the magnetic field B in space.

Also $\nabla^2 W_G = 4\pi EBc\varepsilon \sqrt{A}$ where c is the speed of the electromagnetic wave, A area onto which the fields are perpendicular and ε is the permitivity of free space. Hence the true displacement current is given by

$I = 4\pi\varepsilon AgB$

This equation predicts that either a changing gravitational field or a changing magnetic field produces a current

Gravitational Waves Equation

Using Ampere's circuital law to the equation

$I = \mu g \, [\nabla^2 W_G]^2 / 4\pi E^2 B$

The intensity of the wave is given by

$EB/\mu = \nabla^2 W_G \, (\sqrt{g}/4\pi\Delta l)$

The left hand side of the equation is Maxwell's proof (EB/μ) and the right hand side is a new proof where Δl is the change in the distance moved by a wave. Therefore intensity P_G is related to the potential gradient and the gravity frequency $f_g = (\sqrt{g}/4\pi\Delta l)$.

$P_G = f_g \nabla^2 W_G$

Therefore when we talk of the intensity of the electromagnetic wave we should not forget the frequency of oscillation of particles in the gravitational field.

CHAPTER25: Newton's Biggest Blunder: Re-defining Gravity

It is known that Newton deduced his law of universal gravitation from Keplers laws. What if Kepler made a mistake in his calculations, What if the same mistake is still hidden in Newton's law? What if general relativity failed to determine the same error and instead brought complications? What is the error, what is this dimensionless constant (a number)? Does uncovering this error lead to a complete quantum theory of gravity? Is this error or number the missing link between gravity and quantum mechanics?

Below I re-define gravity in which I rediscover the missing number, a dimensionless physical quantity in Newton's law of gravitation.

Rule 1

The ratio of the velocity of a particle in motion (v) to the speed of light (c) must be eqaul to the ratio of the potential energy of the particle (W_o) to the energy carried by the particle (photon) of light (W)

$$\frac{v}{c} = \frac{W_o}{W}$$

Orbits around the Sun

From rule one a particle will move with a velocity relative to that of light. The radiations or electromagnetic waves emitted by the sun contain changing electric E and magnetic fields B related to the speed of the electromagnetic waves (light) and then rule one becomes

$$\frac{Bv}{E} = \frac{W_o}{W}$$

Since the energy carried by the photon of light is related to the period of revolution of the particle in orbit $\frac{h}{T}$, where h is the Planck constant and the potential energy of the particle in orbit around the sun in the gravitational field of the sun is also related to its mass m and the distance r between the sun and the particle, then the period of revolution of the particle around the sun is

$$T = \frac{Bvhr}{EGM^2}$$

Where G- Universal gravitational constant

It must also be true that the electric field experienced by the particle in orbit around the sun due to the sun's electric force is proportional to the charge q on the particle in orbit but falls with the square of the distance from the sun, and then the period of orbit is

$$T = \frac{4\pi\varepsilon_0 B v h r^3}{qGM}$$

Where, ε_0 - the permittivity of free space

Since particles exhibit wave properties then the velocity of the particle in orbit is related to its wavelength and period of orbit by $v = \frac{\lambda}{T}$, from which the we get the square of the period as

$$T^2 = \frac{4\pi^2}{GM}\left(\frac{2\varepsilon_0 B\lambda\hbar}{Mq}\right)r^3$$

Where $\hbar = \frac{h}{2\pi}$

This is the advanced or modified Kepler's periods (Harmonic Law). The quantity in the brackets $\left(\frac{\varepsilon_0 B\lambda\hbar}{Mq}\right)$ is a dimensionless constant and it is what is missing in Newton's derivation of Kepler's Laws.

Of The Galactic And Atomic

When the dimensionless constant is unity, we have the galactic and atomic descriptions of reality separated. In other words we have to parts of the universe that are separated and need to be merged. The modified Kepler harmonic law given above is a true description of the universe involving all the parts that is, the small and large particles. The galactic is when the dimensionless constant is unity. The atomic is given when

$$\left(\frac{2\varepsilon_0 B\lambda\hbar}{Mq}\right) = 1$$

The wavelength of which is

$$\lambda = \frac{Mq}{2\varepsilon_0 B\hbar}$$

For $B = \frac{E}{v}$ and $E = \frac{q}{4\pi\varepsilon_0 r^2}$ we have

$$\frac{4\pi^2 r^2}{\lambda} = \frac{h}{Mv}$$

Rule 2

The circumference of the smallest allowed orbit for the particle is equal to the wavelength of the particle. This statement is a proof for the quantized energy of the particle in an atom and is expressed as $2\pi r = \lambda$. With this rule, the wavelength of the particle is therefore related to its momentum mv by

$$\lambda = \frac{h}{Mv}$$

It is therefore true according to deBrogile and the derivation given here that particles also have wave properties.

From the derivations given here it is clear that we have a description for both the microscopic and macroscopic particles.

The modified Newton's universal law of gravitation (MONG)

Following the same procedure that Newton used in the derivation of Kepler's harmonic law, the advanced law derived here can be deduced using the advanced Gravitational force given as

$$MONGFORCE = \frac{GMf}{r^2}\left(\frac{Mq}{2\varepsilon_0 B\lambda h}\right)$$

This is arranged as

$$MONGFORCE = \frac{4\pi^2 Gk_e qM}{r^2}\left(\frac{1}{B\lambda h}\right)$$

Where $k_e = \frac{1}{4\pi\varepsilon_0}$ - the coulomb constant

For $2\pi r = \lambda$

$$MONGFORCE = \frac{2\pi Gk_e qM}{r^3}\left(\frac{1}{Bh}\right)$$

The application of the above force is given in chapter11 (see Equation k) where it is used to derive the equation for the entropy of the black hole.

Chapter26: on the origin of gravity, electricity and the laws of Newton

The first attempt into the derivation of the gravitational force and Newton's laws of inertia was given in part by Erik Verlinde (2011) in which he stated that gravity is an entropic force. Simple as it was, his ideas were on a large scale rejected by mainstream physicists. The rejection of verlinde ideas where not backed up by another approach as it has been with other theories save for Sabine Hossenfelder approaches and critics. I think Erik was not surprised by these attacks because this is what physicists do especially when ones analysis or derivation doesn't involve the use of rigorous mathematical models-the one which were used by Einsten and others.

Anyway, what could be another approach towards the derivation of the gravitational, electricity and the law of inertia different from Verlindes idea of the emergent of gravity as an entropic force?

In this chapter we derive Newton's laws of inertia, gravitation and also the electromagnetic force law from first principles without assuming dark matter and the MOND theories. To differ from Verlindes approach we shall use the notions of Quantum vacuum and the Schwinger effect or limit in QED.

In order to understand the physical significance of the derivation to be given herein, we must remember the Schwinger mechanism (Schwinger, 1951) in Quantum Electrodynamics: a strong electric field, greater than a critical value, can create electron-positron pairs from the quantum vacuum.

The Schwinger mechanism has two cornerstones, the first one is the existence of quantum vacuum and the second one the existence of an external electric field (which attempts to separate electrons and positrons). There are no particles in the vacuum (in that sense the vacuum is empty); but the vacuum is plenty of short-living virtual particle-antiparticle pairs which in permanence appear and disappear (allowed by time- energy uncertainty relation). In simple words, the quantum vacuum is a kingdom of the virtual particle-antiparticle pairs; a kingdom with apparently perfect symmetry between virtual matter and virtual antimatter.

A "virtual" pair can be converted into a real electron-positron pair only in the presence of a strong external field, which can spatially separate electrons and positrons, by pushing them in opposite directions, as it does an electric field. Thus, "virtual" pairs are spatially separated and converted into real pairs by the expenditure of the external field energy. For this to become possible, the potential energy has to vary by an amount in the range of about one Compton wavelength , which leads to the conclusion that a significant pair creation occurs only in a very strong external field E, greater than the critical value.

Therefore the external force which attempts to separate particles and antiparticles converting a virtual pair into a real one may be simplified as,

$$F\Phi = 4\pi E_o\hbar \quad \text{(a)}$$

Where F is the external force, Φ is the magnetic flux impeding a sphere of radius R and area $A = 4\pi R^2$ and E_o is the electric field on an electron of mass m (see Chapter9)

Derivation of force and inertia

From chapter 9, the magnetic field is related to the electric field by

$$E_o = B_o c = \frac{Mmec^2}{4\pi\varepsilon_o\hbar^2} \quad \text{(b)}$$

Remember the above given value is constant for a particle with mass m. There is also an assumption for M=m. Note also that $B_o \neq B$

A new assumption and probably the most surprising one is that, the magnetic flux can be related to energy W and acceleration a by

$$\Phi = \frac{We}{a\hbar\varepsilon_o} \quad \text{(c)}$$

Thus, "virtual" pairs are spatially separated and converted into real pairs by the expenditure of the external field energy. For this to become possible, the potential energy has to vary by an amount in the range of about one Compton wavelength , which leads to the conclusion that a significant pair creation occurs only in a very strong external field E, greater than the critical value E_o.

But, $W = Mc^2$, it is therefore evident why equation (c) for the magnetic flux was chosen to be of the given form. It is picked precisely in such a way that one recovers the second law of Newton

F=ma

As easily verified by combining (a) together with (b) and (c).

Therefore, a similarity or resemblance between acceleration, thermodynamics and electromagnetism comes alive in the statement below;

As there is a formula for the temperature T that is required to cause an acceleration equal to a, $T = \frac{ha}{2\pi kc}$ so there must also be a temperature required to cause an acceleration for an electron in the quantum vacuum at a constant electric and magnetic flux, $T = \frac{ha\varepsilon_o\Phi}{ke}$

Derivation of Newton's law of Gravity

Suppose our universe is a sphere of area $A = 4\pi R^2$ with a sea of virtual particles in a quantum vacuum. It is theorized as before that to separate the virtual particle and antiparticles in a vacuum into real particles one will require a strong external electric or magnetic field B. For the gravitational field, this external magnetic field was calculated in chapter 9 to be,

$$B = \frac{ec^2}{4\pi Gh\varepsilon_o}$$

Where we introduced a new constant G. Eventually this constant is going to be identified with Newton's constant, of course. But since we have not assumed anything yet about the existence of a gravitational force, one can simply regard this equation as the definition of G.

Then, the magnetic flux will be given as,

$$\Phi = BA = \frac{ec^2 R^2}{Gh\varepsilon_o}(d)$$

Substituting Equation (b) and (d) into (a) one obtains the familiar law

$$F = G\frac{Mn}{R^2}$$

We have recovered Newton's law of gravitation, practically form first principles. Following the above derivation carefully, it implies that gravity is a quantum force resulting from the quantum fluctuations of the vacuum due to an existence of an external strong electric or magnetic field separating particles from antiparticles or matter from anti matter.

Derivation of the law of electromagnetism

Following the same steps as in the previous derivation for the gravitational force but assuming a different external magnetic field (Schwinger limit as given in chapter9),

$$B = \frac{Mmc^2}{he}$$

Then, the magnetic flux will be given as,

$$\Phi = BA = \frac{4\pi Mmc^2 R^2}{he} \quad (e)$$

Substituting Equation (e) and (b) into (a) one obtains the familiar law

$$F = \frac{e^2}{4\pi \varepsilon_o R^2}$$

We have recovered Coulomb's law of electromagnetism, practically form first principles.

Therefore in principle, every external force which attempts to separate particles and antiparticles, may convert a virtual pair into a real one. If it is always an attractive force, as commonly believed today, gravity can't separate particles and antiparticles. Hence, the conjectured gravitational repulsion between matter and antimatter is a necessary condition for separation of particles and antiparticles by a gravitational field and consequently for the creation of particle-antiparticle pairs from the quantum vacuum. But while an electric field can separate only charged particles, gravitation as a universal interaction might create particle-antiparticle pairs of both charged and neutral particles. Thus, the hypothesis of antigravity opens possibility for a gravitational version of the Schwinger mechanism.

In conclusion, gravity is not an entropic force but rather a quantum force stemming from the quantum fluctuation of particles in the vacuum. That is the main statement, which is new and has not been made before. If true, this should have profound consequences.

Final Notice: A Modification of the Newtonian Dynamics and General relativity in a limit of Small Accelerations in the Presence of the Electric and Magnetic Fields

Hypothesis

From equation (c) Chapter26 we showed that the gravitational acceleration of a particle is related to the kinetic energy W and magnetic flux Φ by the following equation,

$$a_g = \frac{Wq}{\Phi h \varepsilon_0}$$

q-charge on a particle

The above equation implies that, keeping the energy a constant, the acceleration of a particle increases with a fall in the magnetic flux and vice versa is true. Thus the acceleration of a particle is to a great extent affected by the magnetic field.

From the above given relationship we deduce a modification of the Newtonian dynamics in a limit of small accelerations similar to the Milgrom hypothesis (Astrophys. J. 270, 365-1983) but with an electric force as a cause of these small accelerations and we deduce the value of this acceleration in relation to the movement of electrons in the hydrogen atom. Using the same analysis we also deduce the value of the total mass of the universe.

If we let the energy or the kinetic energy, the work done to move the particle around a body of mass M (a galaxy) in the magnetic field created by M, in a process of magnetic induction be given by, $W = mv^2$, where v is the velocity of the particle and m is the mass of the particle. Also if we let the magnetic flux at any point in space at a distance R from M be $\Phi = 4\pi R^2 B$, then the acceleration is simplified in this way,

$$a_g = \frac{mv^2 q}{4\pi R^2 B h \varepsilon_0}$$

This can then be re-written to resemble the Milogram formula in this way,

We know the magnetic field B is related to the velocity and electric field E by, Bv=E, then the acceleration of a particle is,

$$a_g = m\left[\frac{qa}{4\pi\varepsilon_o Ehv}\right]a$$

Where, $a = \frac{v^2}{R}$

This then reduces to,

$$a_g = \left[\frac{q^2 a}{4\pi\varepsilon_o hv\left(\frac{Eq}{m}\right)}\right]a$$

This then gives,

$$a_g = \left[\frac{a}{\frac{1}{a_e}\left(\frac{Eq}{m}\right)}\right]a$$

From the above given given equation, $\alpha_e = \frac{q^2}{4\pi\varepsilon_o hv}$ is the fine structure constant. Eq is the electric force on the particle of charge q. Then the acceleration of a particle of mass m in the electric field will be given by, $a_1 = \frac{Eq}{m}$.

Now relating this to the Milogram Hypothesis of $a_g = \left[\frac{a}{a_o}\right]a$, where $a_o \sim 1.2 \times 10^{-10} m/s^2$, we have

$$\left[\frac{a}{\frac{1}{a_e}a_1}\right]a = \left[\frac{a}{a_o}\right]a$$

Therefore in a limit where v=c (the speed of light) and q=e (the charge on an electron), we have

$$\alpha_e = \frac{e^2}{4\pi\varepsilon_o hc} = 1/137$$

This implies that, the acceleration of a particle in an electric field of a body of mass M will be given as

$$a_1 = \alpha_e a_o = 8.759 \times 10^{-13} m/s^2 \text{ (d)}$$

The above given value implies that, in the presence of the electric and magnetic field the Milogram acceleration constant must be corrected to fit the data very well.

Then the modified Newtonian law of gravity is written as,

$$F = \frac{GMr}{\frac{\alpha_e a}{a_1}R^2}$$

In a limit $a_1 = a$ we don't recover the Newtonian law except for $\alpha_e = 1$, which proves to be difficult. This therefore requires us to rethink gravity with a new gravitational constant that could read as

$$G_B = \frac{G}{\alpha_e} = 9.138 \times 10^{-09} Nm^2/kg^2$$

Calculation of the total mass of the galaxy from equation d

Remember previously in chapter 12 we deduced the ratio of the accelerations to be equal to the finestructure constant as,

$$\frac{g_b}{g_a} = \alpha_e \quad (e)$$

Where,

$$g_a = \frac{8\pi\varepsilon_o hc^5}{e^2 GM}$$

And,

$$g_b = \frac{c^4}{GM}$$

Relating these accelerations to the ones given in equation d we should be able to deduce the mass of the galaxy as,

$$g_a = \frac{8\pi\varepsilon_o hc^5}{e^2 GM} = a_0 = 1.2 \times 10^{-10} m/s^2$$

From which the total galaxy mass is given as,

$$e^2 M = 7.122 \times 10^{18} kgC^2$$

For $e = 1.602 \times 10^{-19}C$, the total mass is given as,

$$M = 5.55 \times 10^{56} kg$$

This is almost the total mass of the Universe.

Note: since the total mass of ordinary matter in the universe is known to be $M = 1.5 \times 10^{53} kg$, *then using the above given equation, the charge required to give this value is* $e = 6.891 \times 10^{-18}C$

But with,

$$g_b = \frac{c^4}{GM} = \alpha_e a_0 = 8.759 \times 10^{-13} m/s^2$$

We have, $M = 1.387 \times 10^{56} kg$

Because this is not the same mass obtained with the use of g_a, we then rewrite the equation of g_b to give the same value of mass as,

$$g_b = \frac{nc^4}{GM}$$

Where, n is the number taking on values from, n=2,4,16,.............

In conclusion, we have not only modified the Newtonian dynamics as a requirement to account for darkmatter but we have just deduced the total mass of the universe. This implies that we are near or probably we have just completed the theory of quantum gravity.

Additional Readings

Balungi Francis, (2010) "A hypothetical investigation into the realm of the microscopic and macroscopic universes beyond the standard model" general physics arXiv:1002.2287v1[1] [physics.gen-ph]

Hawking, Stephen[2] (1975). "Particle Creation by Black Holes"[3]. Commun. Math. Phys.[4] 43 (3): 199–220. Bibcode[5]:1975CMaPh..43..199H[6].

Hawking, S. W.[7] (1974). "Black hole explosions?". Nature.248(5443):30–31.

Bibcode[8]:1974Natur.248...30H[9].doi[10]:10.1038/248030a0[11].

Carlo Rovelli (2003) "Quantum Gravity" Draft of the Book Pdf

Some few texts used are from Wikipedia https://creativecommons.org/licenses/by-sa/3.0/

D. N. Page, Phys. Rev. D 13, 198 (1976).

C. Gao and Y.Lu, Pulsations of a black hole in LQG (2012) arXiv:1706.08009v3

A.H. Chamseddine and V.Mukhanov, Non singular black hole (2016) arXiv 1612.05861v1

M.Bojowald and G.M.Paily, A no-singularity scenario in LQG (2012) arXiv: 1206.5765v1

P.Singh, class.Quant.Grav,26,125005(2009), arXiv:0901.2750

P.Singh and F.Vidotto, Phys.Rev, D83,064027(2011) arXiv:1012.1307

---

1. https://arxiv.org/abs/1002.2287v1

2.   https://en.wikipedia.org/wiki/Stephen_Hawking

3.   http://www.springerlink.com/content/c4553033029k5wk6/

4.   https://en.wikipedia.org/wiki/Commun._Math._Phys.

5.   https://en.wikipedia.org/wiki/Bibcode

6.   http://adsabs.harvard.edu/abs/1975CMaPh..43..199H

7.   https://en.wikipedia.org/wiki/Stephen_Hawking

8.   https://en.wikipedia.org/wiki/Bibcode

9.   http://adsabs.harvard.edu/abs/1974Natur.248...30H

10.   https://en.wikipedia.org/wiki/Digital_object_identifier

11.   https://doi.org/10.1038%2F248030a0

C.Rovelli and F.Vidotto, Phy. Rev,111(9) 091303(2013) arXiv:1307.3228v2

M.Bojowald, Initial conditions for a universe, Gravity Research Foundation (2003)

A.Ashtekar, Singularity Resolution in Loop Quantum Cosmology (2008) arXiv:0812.4703v1

J.Brunneumann and T.Thiemann, On singularity avoidance in Loop Quantum Gravity (2005) arXiv:0505032v1

L.Modesto, Disappearence of the Black hole singularity in Quantum gravity (2004) arXiv:0407097v2

Mikhailov, A.A. (1959).Mon. Not. Roy. Astron. Soc.,119, 593.

P. Merat etal.(1974). Astron & Astrophys 32, 471-475

Trempler, R.J. (1956).Helv. Phys. Acta, Suppl.,IV, 106.

Trempler, R.J. (1932). " The deflection of light in the sun's gravitational field "Astronomical society of the pacific 167

Einstein, A. (1916).Ann. d. Phys.,49, 769; (1923).The Principle of Relativity, (translators Perret, W. and Jeffery, G.B.), (Dover Publications, Inc., New York), pp. 109–164.

Von Klüber, H. (1960). InVistas in Astronomy, Vol. 3, pp. 47–77.

K. Hentschel (1992). Erwin Finlay Freundlich and testing Einstein theory of relativity, Communicated by J.D. North

Muhleman, D.O., Ekers, R.D. and Fomalont, E.B. (1970).Phys. Rev. Lett.,24, 1377

Mikhailov, A.A. (1956).Astron. Zh.,33, 912.

Dyson, F.W., Eddington, A.S. and Davidson, C. (1920).Phil. Trans. Roy. sog., A220, 291

Chant, C.A. and Young, R.K. (1924).Publ. Dom. Astron. Obs.,2, 275.

Campbell, W.W. and Trumbler, R.J. (1923).Lick Obs. Bull.,11, 41.

Freundlich, E.F., von Klüber, H. and von Brunn, A. (1931).Abhandl. Preuss. Akad. Wiss. Berlin, Phys. Math. Kl., No.l;Z. Astrophys.,3, 171

Mikhailov, A.A. (1949).Expeditions to Observe the Total Solar Eclipse of 21 September, 1941, (report), (ed. Fesenkov, V.G.), (Publications of the Academy of Sciences, U.S.S.R.), pp. 337–351.

S.P. Martin, in Perspectives on Supersymmetry , edited by G.L. Kane (World Scientific, Singapore, 1998) pp. 1–98; and a longer archive version in hep-ph/9709356; I.J.R. Aitchison, hep-ph/0505105.

Mara Beller, Quantum Dialogue: The Making of a Revolution. University of Chicago Press, Chicago, 2001.

Morrison, Philp: "The Neutrino, scientific American, Vol 194,no.1 (1956),pp.58-68.

R. Haag, J. T. Lopuszanski and M. Sohnius, Nucl. Phys. B88, 257 (1975) S.R. Coleman and J. Mandula, Phys.Rev. 159 (1967) 1251.

H.P. Nilles, Phys. Reports 110, 1 (1984).

P. Nath, R. Arnowitt, and A.H. Chamseddine, Applied N = 1 Supergravity (World Scientific, Singapore, 1984).

S.P. Martin, in Perspectives on Supersymmetry , edited by G.L. Kane (World Scientific, Singapore, 1998) pp. 1–98; and a longer archive version in hep-ph/9709356; I.J.R. Aitchison, hep-ph/0505105.

S. Weinberg, The Quantum Theory of Fields, VolumeIII: Supersymmetry (Cambridge University Press, Cambridge,UK, 2000).

E. Witten, Nucl. Phys. B188, 513 (1981).

S. Dimopoulos and H. Georgi, Nucl. Phys. B193, 150(1981).

N. Sakai, Z. Phys. C11, 153 (1981);R.K. Kaul, Phys. Lett. 109B, 19 (1982).

L. Susskind, Phys. Reports 104, 181 (1984).

L. Girardello and M. Grisaru, Nucl. Phys. B194, 65(1982); L.J. Hall and L. Randall,

Phys. Rev. Lett. 65, 2939(1990);I. Jack and D.R.T. Jones, Phys. Lett. B457, 101 (1999).

For a review, see N. Polonsky, Supersymmetry: Structureand phenomena. Extensions of the standard model, Lect.Notes Phys. M68, 1 (2001).

G. Bertone, D. Hooper and J. Silk, Phys. Reports 405, 279 (2005).

G. Jungman, M. Kamionkowski, and K. Griest, Phys. Reports 267, 195 (1996).

V. Agrawal, S.M. Barr, J.F. Donoghue and D. Seckel,Phys. Rev. D57, 5480 (1998).

N. Arkani-Hamed and S. Dimopoulos, JHEP 0506, 073(2005); G.F. Giudice and A. Romanino, Nucl. Phys. B699, 65(2004) [erratum: B706, 65 (2005)]. July 27, 2006 11:28

en.wikipedia.org/wiki/Supersymmetry - 52k - Cached[12] - Similar pages[13]

en.wikipedia.org/wiki/Grand_unification_theory - 39k - Cached[14] - Similar pages[15]

In cosmology, the Planck epoch (or Planck era), named after Max Planck, is the earliest period of time in the history of the universe, ...

en.wikipedia.org/wiki/**Planck_epoch** - 23k - Cached[16] - Similar pages

L. Shapiro and J. Sol`a, Phys. Lett. B 530, 10 (2002);

E. V.Gorbar and I. L. Shapiro, JHEP 02, 021 (2003); A. M. Pelinson,

L. Shapiro, and F. I. Takakura, Nucl. Phys. B 648, 417 (2003).

A. Starobinsky, Phys. Lett. B 91, 99 (1980).

G. F. R. Ellis, J. Murugan, and C. G. Tsagas, Class. Quant. Grav.21, 233 (2004).

H. V. Peiris et al., Astrophys. J. Suppl. 148, 213 (2003).

D. N. Spergel et al., astro-ph/0603449.

Vilenkin, Phys. Rev. D 32, 2511 (1985).

A. Starobinsky, Pis'ma Astron. Zh 9, 579 (1983).

A.H. Guth, Phys. Rev. D23, 347 (1981).

A.D. Linde, Phys. Lett. B108, 389 (1982); A. Albrecht, P.J. Steinhardt, Phys.Rev. Lett. 48, 1220 (1982).

A.D. Linde, Phys Lett. B129, 177 (1983).

N. Makino, M. Sasaki, Prog. Theor. Phys. 86, 103 (1991); D. Kaiser, Phys. Rev.D52, 4295 (1995).

H. Goldberg, Phys. Rev. Lett. 50, 1419 (1983).

---

12.   http://64.233.169.104/search?q=cache:ZBSZWNLrdxEJ:en.wikipedia.org/wiki/
        Supersymmetry+supersymmetry&hl=en&ct=clnk&cd=1&gl=ug

13.   http://www.google.co.ug/search?hl=en&q=related:en.wikipedia.org/wiki/Supersymmetry

14.   http://64.233.169.104/search?q=cache:Le3KZW7QnUYJ:en.wikipedia.org/wiki/
        Grand_unification_theory+grand+unification&hl=en&ct=clnk&cd=1&gl=ug

15.   http://www.google.co.ug/search?hl=en&q=related:en.wikipedia.org/wiki/Grand_unification_theory

16.   http://64.233.169.104/search?q=cache:d5zWrem4T08J:en.wikipedia.org/wiki/
        Planck_epoch+planck+epoch&hl=en&ct=clnk&cd=1

E. Kolb and M. Turner, The Early Universe (Addison-Wesley, Reading, MA, 1990).

W. Garretson and E. Carlson, Phys. Lett. B 315, 232(1993); H. Goldberg, hep-ph/0003197.

Eddington, A. S., The Internal Constitution of the Stars (Cambridge University Press, England, 1926), p. 16

Duncan R .C. & Thompson C., Ap.J.392, L 9 (1992).

Thompson , C, Duncan , R.C ., Woods , P., Kouveliotou , C ., Finger , M.H. & van Parad ij s , J .,ApJ, submitted , astro-ph /9908086, (2000).

Schwinger , J .,Phys. Rev.73, 416L (1948)

Carlip, S.: Quantum gravity: a progress report. Rept. Prog. Phys. 64, 885 (2001).arXiv:gr-qc/0108040

Kerr,R.P.: Gravitational field of a spinning mass as an example of algebraically special metrics.

Phys. Rev. Lett. 11, 237–238 (1963)

Bekenstein, J.: Black holes and the second law. Lett. Nuovo Cim. 4, 737–740 (1972)

Bardeen, J.M., Carter, B., Hawking, S.: The four laws of black hole mechanics. Commun.

Math. Phys. 31, 161–170 (1973)

Tolman, R.: Relativity, Thermodynamics, and Cosmology. Dover Books on Physics Series.

Dover Publications, New York (1987)

Oppenheimer, J., Volkoff, G.: On massive neutron cores. Phys. Rev. 55, 374–381 (1939)

Tolman, R.C.: Static solutions of einstein's field equations for spheres of fluid, Phys. Rev. 55,

364–373 (1939)

Zel'dovich Y.B.: Zh. Eksp. Teoret. Fiz.41, 1609 (1961)

Bondi, H.: Massive spheres in general relativity. Proc. Roy. Soc. Lond. A281, 303–317 (1964)

Sorkin, R.D., Wald, R.M., Zhang, Z.J.: Entropy of selfgravitating radiation. Gen. Rel. Grav.

1127–1146 (1981)

Newman, E.T., Couch, R., Chinnapared, K., Exton, A., Prakash, A., et al.: Metric of a rotating,

charged mass. J. Math. Phys. 6, 918–919 (1965)

Ginzburg, V., Ozernoi, L.: Sov. Phys. JETP 20, 689 (1965)

Doroshkevich, A., Zel'dovich, Y., Novikov I.: Gravitational collapse of nonsymmetric and rotating masses, JETP 49 (1965)

Israel, W.: Event horizons in static vacuum space-times. Phys. Rev. 164, 1776–1779 (1967)

Israel,W.: Event horizons in static electrovac space-times. Commun. Math. Phys. 8, 245–260 (1968)

Loop quantum gravity does provide such a prediction [363, 364], and it disagrees with the semiclassical

Carter, B.: Axisymmetric black hole has only two degrees of freedom. Phys. Rev. Lett. 26, 331–333(1971)

Penrose, R.: Gravitational collapse: the role of general relativity. Riv. Nuovo Cim. 1, 252–276 (1969)

Christodoulou, D.: Reversible and irreversible transformations in black hole physics. Phys. Rev. Lett. 25, 1596–1597 (1970)

Christodoulou, D., Ruffini, R.: Reversible transformations of a charged black hole. Phys. Rev. D4, 3552–3555 (1971)

Hawking, S.: Particle creation by black holes. Commun. Math. Phys. 43, 199–220 (1975)

Klein, O.: Die reflexion von elektronen an einem potential sprung nach der relativistischen dynamik von dirac. Z. Phys. 53, 157 (1929)

Wald, R.M.: General Relativity. The University of Chicago Press, Chicago (1984)

Hawking, S.W.: Black hole explosions. Nature 248, 30–31 (1974)

Hawking, S., Ellis, G.: The large scale structure of space-time. Cambridge University Press, Cambridge (1973)

Carter, B.: Black hole equilibrium states, In Black Holes—Les astres occlus. Gordon and Breach Science Publishers, (1973)

Hawking, S.W.: Gravitational radiation from colliding black holes. Phys. Rev. Lett. 26, 1344– 1346 (1971)

Hawking, S.: Black holes in general relativity. Commun. Math. Phys. 25, 152–166 (1972)

Bekenstein, J.: Extraction of energy and charge from a black hole. Phys. Rev. D7, 949–953 (1973)

Bekenstein, J.D.: Black holes and entropy. Phys. Rev. D7, 2333–2346 (1973)

Hawking, S.: Quantum gravity and path integrals. Phys. Rev. D18, 1747–1753 (1978)

Gross, D.J., Perry, M.J., Yaffe, L.G.: Instability of flat space at finite temperature. Phys. Rev. D25, 330–355 (1982)

Unruh, W.G., Wald, R.M.: What happens when an accelerating observer detects a rindler particle. Phys. Rev. D29, 1047–1056 (1984)

Bekenstein, J.D.: Auniversal upper bound on the entropy to energy ratio for bounded systems.

Phys. Rev. D23, 287 (1981)

Unruh,W.,Wald, R.M.: Acceleration radiation and generalized second law of thermodynamics. Phys. Rev. D25, 942–958 (1982)

Unruh, W., Wald, R.M.: Entropy bounds, acceleration radiation, and the generalized second law. Phys. Rev. D27, 2271–2276 (1983)

Image : MPI for gravitational physics / W.Benger-zib

Tomilin,K.A., (1999). "Natural Systems Of Units: To The Centenary Aniniversary Of The Planck Systems", 287-296

Sivaram, C. (2007). "What Is Special About the Planck Mass"? arXiv:0707.0058v1

H. Georgi and S.L. Glahow. (1974) "Unity Of All Elementary-Particle Forces". Phys. Rev. Letters 32, 438

Connect With Balungi Francis

I really appreciate you reading my book

If you enjoyed this book or found it useful I'd be very grateful if you'd post a short review. Your support really does make a difference and I read all the reviews personally so I can get your feedback and make this book even better.

Friend me on Facebook: http://facebook.com/balungi.francis[17]

Visit my Facebook page Visionary School of Quantum Gravity: http://facebook.com/BalungiF[18]

Subscribe to my blog: http://visionaryschoolofquantumgravity.blogspot.ug[19]

Contacts,

Tel: +256(0)777105605

+256(0)703683756

Email: balungif@gmail.com

Thanks again for your support!

---

17. http://facebook.com/markcoker

18. http://facebook.com/markcoker

19. http://blog.smashwords.com

Don't miss out!

Visit the website below and you can sign up to receive emails whenever Balungi Francis publishes a new book. There's no charge and no obligation.

https://books2read.com/r/B-A-LFLG-GEWT

BOOKS 2 READ

Connecting independent readers to independent writers.

Also by Balungi Francis

Beyond Einstein
Quantum Gravity in a Nutshell1
Solutions to the Unsolved Physics Problems
Mathematical Foundation of the Quantum Theory of Gravity
A New Approach to Quantum Gravity
Balungi's Approach to Quantum Gravity
QG: The strange theory of Space,Time and Matter
The Holy Grail of Modern Physics
Fifty Formulas that Changed the World
Quantum Gravity in a Nutshell1 Second Edition
What is Real?:Space Time Singularities or Quantum Black Holes?Dark
Matter or Planck Mass Particles? General Relativity or Quantum Gravity?
Volume or Area Entropy Law?
The Holy Grail of Modern Physics
Brief Solutions to the Big Problems in Physics, Astrophysics and Cosmology

Brief Solutions to the Big Problems
Brief Solutions to the Big Problems

Pursuing a Nobel Prize
Serious Scientific Answers to Millennium Physics Questions

Using Geographical Information Systems to Create an Agroclimatic Zone map for Soroti District

Think Physics
Proof of the Proton Radius
Emergence of Gravity
On the Deflection of Light in the Sun's Gravitational Field
Reinventing Gravity

Standalone
Using Gis to Create an Agroclimatic Zone map for Soroti Distric
Expecting
Quantum Gravity in a Nutshell 2
Balungi's Guide to a Healthy Pregnancy
Prove Physics
The Origin of Gravity and the Laws of Physics
Derivation of Newton's Law of Gravitation
When Gravity Breaks Down